Uwe M. Borghoff

Catalogue of Distributed File/Operating Systems

Springer-Verlag

Berlin Heidelberg New York
London Paris Tokyo
Hong Kong Barcelona
Budapest

Dr. Uwe M. Borghoff

Institut für Informatik, Technische Universität München
Postfach 20 24 20, W-8000 München 2, FRG

ISBN-13: 978-3-540-54450-0   e-ISBN-13: 978-3-642-76880-4
DOI: 10.1007/978-3-642-76880-4

Camera ready by author

45/3140-5 4 3 2 1 0 - Printed on acid-free paper

In general, distributed systems can be classified into Distributed File Systems (DFS) and Distributed Operating Systems (DOS). The survey which follows distinguishes between DFS approaches in Chapters 2–3, and DOS approaches in Chapters 4–5. Within DFS and DOS, I further distinguish "traditional" and object-oriented approaches. A traditional approach is one where processes are the active components in the systems and where the name space is hierarchically organized. In a centralized environment, UNIX would be a good example of a traditional approach. On the other hand, an object-oriented approach deals with objects in which all information is encapsulated. Some systems of importance do not fit into the DFS/DOS classification. I call these systems "closely related" and put them into Chapter 6. Chapter 7 contains a table of comparison. This table gives a lucid overview summarizing the information provided and allowing for quick access. The last chapter is added for the sake of completeness. It contains very brief descriptions of other related systems. These systems are of minor interest or do not provide transparency at all. Sometimes I had to assign a system to this chapter simply for lack of adequate information about it.

This book was influenced by many people, first of all by the designers of the surveyed systems. It is impossible to acknowledge all of them. I want especially to thank Hyo Ashihara, Jean Bacon, Michel Banatre, Yolande Berbers, Ken Birman, Mark Brown, Peter Buhler, Luis-Felipe Cabrera, David Cheriton, Partha Dasgupta, Fred Douglis, Jan Edler, Brett Fleisch, David Gifford, Ray Giguette, Andrzej Goscinski, Marc Guillemont, Robert Hagmann, Hermann Härtig, Ralf-Guido Herrtwich, Norman Hutchinson, Lori Iannamico, Douglas Jensen, Oliver Kowalski, Sacha Krakowiak, Hank Levy, Barbara Liskov, Ami Litman, Klaus-Peter Löhr, Brian Marsh, Henry Mensch, Nathaniel Mishkin, Jeffrey Mogul, Daniel Mosse, Kevin Murray, Jürgen Nehmer, Lutz Nentwig, Clifford Neuman, John Nicol, Jerre Noe, David Notkin, Dave Presotto, Calton Pu, Isabelle Puaut, Marc Pucci, Krithi Ramamritham, Brian Randell, Mahadev Satyanarayanan, Rick Schlichting, Wolfgang Schröder-Preikschat, Michael Scott, Marc Shapiro, Alfred Spector, Robert Stroud, Andrew Tanenbaum, Walter Tichy, Anand Tripathi, Thomas Truscott, Andy Wellings, Jeannette Wing, for giving me insight into their systems. Their collective assistance has greatly improved the book by keeping it honest and up-to-date. I also thank Kathy Schlag and the English Copy-Editor J. Andrew Ross of Springer-Verlag who gave very helpful comments on the language.

All remaining omissions, faults and misinterpretations are my own, of course. I apologize in advance for anything that might irritate you. Due to the explosive nature of this subject, it is possible that contact addresses may have changed, or that latest systems are missing. I admit defeat in surveying all transparent distributed systems. I apologize for that as well.

Munich, June 1991   Uwe M. Borghoff

Preface

During the last decade, distributed systems have become a major focus of attention. Due to the development of local area networks (LANs) and metropolitan or wide area networks (MANs or WANs), more and more people are involved in distributed systems. Their interest concerns, for example, transparent distributed computation, transparent access to geographically dispersed data, or the amenity of a higher availability of data if it is replicated among the nodes in the network. Furthermore, they want to know which existing system can provide such features.

This book is intended for people in universities and industry interested in short descriptions of a large number of existing transparent distributed systems.

The basic idea of writing such a book arose after a course on distributed operating systems under the supervision of Hans-Jürgen Siegert. Ernst Biersack and Kristof Nast-Kolb encouraged me to write down brief descriptions of distributed file/operating systems. Kristof Nast-Kolb co-authored an earlier version of this survey [1].

The wonderful surveys by Andrew Tanenbaum and Robbert van Renesse on distributed operating systems [2] and by Liba Svobodova on file servers for network-based distributed systems [3] gave me hints of what such a survey could look like. But surveying more than a hundred systems is quite a job. Therefore, I decided to publish an encyclopedia-style survey where each system is independently described. The description of all systems, however, accords with a catalogue of criteria.

The book contains eight chapters organized as follows. Chapter 1 contains an introduction where I present the way the systems are described. Here definitions are offered of the features of each transparent distributed system as well as the services it is able to provide. Many of the distributed systems provide varying levels of transparency. Therefore, I develop a catalogue of criteria to aid the organization of the survey. Such a scheme allows an implementation-independent description of the systems surveyed. The main entries concern heterogeneity of the system's environment, communication strategy, and naming and security issues. Finally, I examine the reliability and availability of the separate systems and the way these issues are achieved.

Each system is described in stages. First, I introduce the main goal the system was developed for, the classification of the system, and the transparency levels provided. Second, I try to give a short view of the advantages and disadvantages of the system. This allows the reader to compare the systems with respect to trade-off issues, such as performance versus availability. Third, a description of the system is given according to the catalogue of criteria. Finally, I append information on other issues, the current status of the research, whom to contact, and a long list of references.

Contents

Chapter 1

Introduction

This book is yet another survey of distributed systems.[1] However, I believe that this approach is novel. For one thing, I have tried, to the best of my knowledge, to include the majority of the distributed file and operating systems presently in existence or of historical interest. For another, the approach is encyclopedic. Each system is described independently. A description of both the main goal and the pros and cons (advantages and disadvantages) of each system enables the reader to decide whether or not a system is of interest to him or her.

Finally, a table comparing all systems presented is added as well as a long list of references. Nobody needs to read through the whole book if he or she is only interested in some of the points.

1.1 Presentation of a Catalogue of Criteria

I have tried to develop a scheme – referred to as a *catalogue of criteria* – that allows a description of the systems in an implementation-independent way. The main questions to be answered are: What kind of transparency levels are provided, how is each kind of transparency achieved, what kind of communication strategy has been proposed and finally, does the distributed character of the system allow increased availability and reliability. The last question leads on to an analysis of replication schemes used and to an evaluation of proposed failure handling and recovery strategies.

Transparency Levels

The main goal of distributed file systems (DFS) or distributed operating systems (DOS) is to provide some level of transparency to the users of a computer network. I distinguish five levels of transparency. *Location* transparency exists when a process requesting a particular network resource does not necessarily know where the resource is located. *Access* transparency gives a user access to both local and remotely located resources in the same way.

---

[1] Other surveys of importance are [2] (CDCS, Amoeba, V, Eden), [3] (WFS, XDFS, SWALLOW, CFS, Felix, CMCFS), [4] (Newcastle, LOCUS, NFS, Sprite, Andrew), [5] (NFS, RFS, SMB), [6] (NFS, RFS, AFS), [7] (NFS, RFS), [8] (DOMAIN, Newcastle, NFS, LOCUS), [9] (Amoeba, Sprite), [10] (XDFS, CFS), and [11] (Amoeba, Argus, CHORUS, Clouds, Eden, Emerald, Camelot).

For reasons of availability, resources are sometimes replicated. If a user does not know whether a resource has been replicated or not, *replication* transparency exists.

The problem of synchronization is well-known. In a distributed environment this problem arises in an extended form. Encapsulating concurrency control inside a proposed system is what is meant by *concurrency* transparency. This includes schemes that provide system-wide consistency as well as weaker schemes in which user interaction can be necessary to recreate consistency. The distinction between transaction strategies (two-phase commit protocols, etc.) and pure semaphore-based techniques is introduced in a special evaluation.

The last level of transparency is *failure* transparency. Network link failures or node crashes are always present within a network of independent nodes. Systems that provide stable storage, failure recovery and some state information are said to be failure transparent.

Heterogeneity

This survey also gives information on the *heterogeneity* of the systems, i.e., assumptions made about the hardware and the operating system interface. Furthermore, I discuss whether the operating system is a modified or look-alike version of another operating system. I describe the underlying network, and also discuss if it is a LAN, what kind of LAN it is, and whether gateways have been developed to integrate the systems into a WAN world.

Changes Made

There are two main forms of implementing distribution. Either a *new layer* can be inserted on top of an existing operating system that handles requests and provides remote access as well as some of the transparency levels, or the distributed system can be implemented as a *new kernel* that runs on every node. This distinction is a first hint of how portable or compatible a system is. Some systems do not distribute all kernel facilities to all nodes. Dedicated servers can be introduced (strict client/server model). Some systems distribute a small kernel to all nodes and the rest of the utilities to special nodes (non-strict client/server model). Another group of systems are the so-called *integrated* systems. In an integrated system each node can be a client, a server or both. This survey is intended to describe these differences.

Communication Protocols

Message passing is the main form of *communication* (excepting multiprocessor systems which can use shared memory). I show which kind of protocols are used and describe specialized protocols if implemented.

Connection and RPC Facility

The kind of *connection* established by the (peer) communication partners is another important criteria. I distinguish between point-to-point connections (virtual circuits), datagram-style connections, and connections based on pipes or streams. If a remote procedure call (RPC) facility is provided, this information is added as well.

Semantics

The users of a distributed system are interested in the way their services are provided and what their *semantics* are. I distinguish *may-be* semantics (which means that the system guarantees nothing), *at-least-once* semantics (retrying to fulfill a service until acknowledged, sometimes done twice or more frequently), *at-most-once* semantics (mostly achieved by duplicate detection) and *exactly-once* semantics. Exactly-once means exactly one execution in the absence of failures, and at most one in their presence. This is typically achieved by making a service an atomic issue (the so-called all-or-nothing principle).

Naming Strategy

I describe the *naming* philosophy and distinguish between *object-oriented* and *traditional* hierarchical naming conventions. The survey includes the proposed name space itself as well as the mechanisms used to provide a system-spanning name space, for example, mounting facilities or superroot-approaches.

Security Issue

Security plays an important role within distributed systems, since the administration could possibly be decentralized and participating hosts cannot necessarily be trusted. Intruders may find it easy to penetrate a distributed environment. Therefore, sophisticated algorithms for encryption and authentication are necessary. I add four entries concerning this issue. First, *encryption* is used if no plain text is exchanged over the communication media. Second, some systems make use of *special hardware* components to achieve security during the message transfer. Third, *capabilities* are provided that enable particular users access to resources in a secure and predefined way. Finally, the entry *mutual authentication* is introduced. This feature is provided if a sort of hand-shake mechanism is implemented that allows bilateral recognition of trustworthiness.

Availability

Distributed systems can be made highly available by replicating resources or services among the nodes of the network. Thus, individual indispositions of nodes can be masked.

(Nested) transactions are well-suited in a computer network. The survey covers this feature. First of all, I look at the concurrency control scheme (optimistic or pessimistic), i.e., *availability* is introduced through the following mechanisms: *synchronization* scheme, *(nested) transaction* facility, and *replication*.

Failure Handling

Failure handling and recovery is a very critical issue. Since some systems are designed to perform well in an academic environment and some systems are made highly reliable for commercial use, trade-off decisions must be taken into account. I add the following entries to the catalogue of criteria. Does the system provide recovery after a *client* or a *server crash*, does it support *orphan detection* and *deletion*, and finally, is there non-volatile memory called *stable storage*.

Process Migration

My final point of interest is *mobility*. Some object-oriented systems provide mobile objects, some traditional process-based systems support migration of processes. Sometimes these approaches come along with load-balancing schemes to increase the system's performance. This issue is included in the survey.

1.2   Organization of the Survey

The following five chapters contain the survey of a large number of distributed systems. Chapter 2 contains a survey of traditional distributed file systems. The survey of object-oriented distributed file systems is given in Chapter 3. Chapters 4 and 5 contain surveys of traditional and object-oriented distributed operating systems respectively. In Chapter 6 I survey closely related systems of high importance which do not fit into the DFS/DOS-classification.

Within each chapter, the systems are described in alphabetical order. Each system's overview is identically structured and described in stages. First, I describe the *main goal* for which the system was developed and add a brief description of the levels of transparency provided. I also give some information on the character of the project, for example, whether it is an academic or a commercial project.

Next, I try to describe the the pros and cons of the system. Knowing both the main goal and the *advantages* will sometimes suffice in deciding whether a particular system is interesting or not. If it is of interest, I include a *description* of the system in question containing all the information needed to understand whether and how the entries of the catalogue of criteria were achieved.

Finally, I append *miscellaneous* information, such as other issues of the system including applications or special software implemented to support the approach. I describe the current status of the project and whom to contact to get more information, and a more or less complete list of references to help the reader in his or her investigation.

The survey ends with a table of comparison (Chapter 7). Everybody knows that in certain ways a categorized comparison can be misleading. Therefore I recommend careful use of this table. It summarizes the main entries of the catalogue of criteria and provides for every system the degree to which it fulfills them.

Chapter 8 contains a list of related projects which are not surveyed in the main survey-chapters because they do not provide the necessary transparency levels or because they are of minor interest, for example, predecessor versions of a surveyed system. They are mentioned for the sake of completeness. Sometimes I had to add a system of importance to this chapter owing to the lack of adequate information about it. Anyway, a brief description and at least a main reference is given for them.

Disclaimer

Dear reader, due to the nature of this book, I know that an up-to-date status cannot be reached. There is a good chance that new publications on the surveyed systems or altogether new systems are not mentioned. Moreover, contact addresses may have changed.

I do not mean to provoke a "religious war" on what is essential for a system to be classified as an object-oriented or a traditional approach. The classification of the surveyed systems may occasionally seem arbitrary. However, my classification tries to correspond with the relevant literature and the designer's declarations and simply aims to give a better overview than a flat organization.

Finally, this book is not a text book. Read it as a reference book or at least as a long bibliography with some "additional" information. But most importantly, enjoy.

Chapter 2

Traditional Distributed File Systems

2.1 Alpine

Main Goal

The primary purpose of Alpine (known within Xerox as *Research Alpine* or *rAlpine*) is to store database files. Its secondary goal is to provide transparent access to ordinary files, such as documents, programs, and the like.

Advantages

The database system at Xerox runs on top of the Alpine transaction-based hierarchical file system. The clear disadvantage of file-level transactions is low performance since a file system sets locks on files or pages. In contrast, a database system is able to perform its own concurrency control and to set locks on logical units. Alpine gives the client no way to find out why a transaction has aborted.

On the other hand, an advantage of the above philosophy of file-level transaction is that it provides the option of running the database system on a separate machine from the file system.

Description

Alpine is written in the Cedar language using the Cedar programming environment (Sect. 2.3). Alpine runs on Dorado workstations.

All communication is performed via a (Cedar) general-purpose Remote Procedure Call facility. A compiler, called *Lupine*, takes a Cedar interface as input and produces server and client stub modules as output. The client stub module exports the Cedar interface access transparently, translating local calls into a suitable sequence of packets addressed to the server.

A client must call the RPC system to establish a *conversation* before it can make authenticated calls. The conversation embodies the authenticated identities of both the client and the server, and the security level (encrypted or not) of their communication.

A file is named by a file-Id that is unique within a so-called *volume* containing the file. A volume is an abstraction of a physical disk volume. Volumes are named and have globally unique Ids.

Alpine uses a *(redo) log*-based technique to implement atomic file update, and performs the standard two-phase commit protocol. The locking hierarchy has two levels: entire-file and page locks. A client chooses between these two levels when opening a file. The log is represented as a fixed-length file in the underlying file system, divided into variable length log records, allocated in a queue-like fashion. It implements a simple access control scheme with a read and modify access list for each file. The Alpine file system supports atomic transactions. Alpine treats a *transaction-Id* as a capability. Any client that has a transaction's Id and the name of the transaction's coordinator may participate in this particular transaction. Clients participating in the same transaction are responsible for synchronization since locks do not synchronize concurrent procedure calls for the same transaction.

A *lock manager* detects deadlocks within a single file system by timeouts and aborts transactions to break these deadlocks.

Replication of files is not supported.

Miscellaneous

OTHER ISSUES: Xerox conceived of building Alpine in December 1980. Three external components have been built to make Alpine usable from Cedar: An IFS-like (Sect. 8.36) hierarchical directory system, an implementation of Cedar open file, and a user interface to the server's administrative functions.

STATUS: By early 1983 the implementation was coded and testing was started. Xerox decided to replace XDFS (Sect. 2.13) by Alpine. Alpine and IFS are both still in operation.

It is planned to include a distributed directory system with location transparency and some form of file replication.

CONTACT: Mark R. Brown, Digital Equipment Corporation, Systems Research Center, 130 Lytton Ave., Palo Alto CA 94301, USA
or Karen N. Kolling, Digital Equipment Corporation, Workstation Systems Engineering, 100 Hamilton Ave., Palo Alto CA 94301, USA
or E.A. Taft, Adobe Systems Inc., 1870 Embarcadero Rd., Suite 100, Palo Alto CA 94303, USA
e-mail: mbrown@decwrl.dec.com or kolling@decwrl.dec.com
REFERENCES: [16], [17]

2.2   Andrew File System (AFS)

Main Goal

The Andrew File System (AFS) was initially developed at Carnegie Mellon University, Pittsburgh, Pennsylvania, beginning in 1982. It is a system providing a data-sharing model of computation (data migrates to the site of computation) built using client/server computing primitives; e.g., threading, name service, RPC. The initial goals this file system were to support at least 7000 workstations on a campus-wide network while providing students, faculty members, and staff with the amenities of a shared file system. AFS now provides a single, location-transparent name space that can span institutional boundaries. Networks of 56kB and above are used to connect "cells", or administrative domains. As of September 1990, AFS has been installed at about 40 sites. The Carnegie Mellon installation is the largest with over 12 000 accounts scattered on nearly 30 file servers used by over 2 000 client machines. AFS 3.0 has been commercialized by Transarc Corporation (Pittsburgh). In May 1990, AFS 4.0 was accepted by the Open Software Foundation as the basis for the data sharing component of its Distributed Computing Environment (DCE).

AFS 2.0 has also been the starting point of additional research via the Coda Project at Carnegie Mellon which is developing a highly-available distributed file system. Coda combines read-write replication of data at servers with the ability for clients to operate totally disconnected for extended periods of time. Disconnected operation is particularly valuable in supporting portable computers. See the Coda section for more detail (Sect. 2.4).

Advantages

Andrew shows that connecting large numbers of heterogeneous workstations is tractable. (for example, there are about 5000 machines that have access to a uniform file name space as of September 1990.) Extensive use of replication is the primary reason for the success of the file system. Replication, via R/O replication of slowly changing user data, consistent distributed replication of system meta-data (for example, the location of volumes, now filesets), and local caching of more rapidly changing data is the primary reason for the success of AFS. Because client machines cache large chunks of data (64 kB when they have disks), both server and network loads are substantially reduced over other file systems such as NFS (Sect. 2.8) or RFS (Sect. 2.9).

Description

AFS code runs on a number of commercial UNIX variants on SUN 3/4 hardware, Digital DECstation/VAXstations, HP 3000, NeXT, IBM RT PCs, and IBM RS/6000s. AFS also runs on MACH (Sect. 4.13). In AFS 3 and 4, a cache manager (which is virtual file system using the VFS interface) intercepts requests for remotely stored files and either obtains the requested data from the cache, or requests the appropriate chunk from the appropriate file server. In AFS 3.0, file servers are typically dedicated in function, but in AFS 4.0, machines may effectively be both clients and servers. To provide backward compatibility, AFS can export its data via NFS protocol.

All machines using AFS can refer to any file stored in AFS using a common name: e.g., in the AFS 3.0 system, the pathname /afs/athena.mit.edu/user/a/xyz can be used anywhere within the AFS system. In AFS 4.0, both Digital's DNS and X.500 are used to navigate through the top-most directories of the name space.

Communication between clients and servers is via RPC. The NCS 2.0 RPC of Hewlett-Packard is the communication vehicle for AFS 4.0, but numerous other RPC mechanisms have been used in previous releases. Typically the RPCs have run using datagram services, but NCS 2.0 has been modified by DEC to run also on connection-oriented OSI TP4.

Since the mid-1980s AFS has had good security due to its use of Needham-Schroeder authentication and access control lists. Early versions of AFS used a CMU-developed authentication mechanism. AFS 3.0 has ACLs on a per-directory basis and used Kerberos V.4 from MIT (Sect. 6.1). AFS 4.0 has POSIX-like ACLs on a per-file basis and uses Kerberos V.5 authentication protocols. Security is administered on a per-cell basis. Within AFS (and now the OSF DCE as well), a cell is the unit of autonomy or administrative control.

Any subtree can be made into a fileset in AFS. Administrators holding appropriate rights can use simple commands to move filesets between servers. The location of filesets is stored within the fileset location database, a database replicated using a transactional commit protocol. Fileset location, like almost all meta-data, is cached on clients to reduce the need to access location data across the network.

With AFS 4.0, POSIX access semantics are provided, with the exception of the precise maintenance of last read time. This is done via a token passing scheme, whereby a token manager precisely manages who has read and write access to particular chunks of data. Token management permits even precise UNIX read/write semantics for mapped files.

Availability in AFS is provided by the replication of all system meta-data, and the replication of read-mostly data. AFS 4.0 adds automatic update of read-only replicas and a fast-restart physical file system, that reduces down-times upon server crashes. The Coda research project substantially aids availability by allowing continued access to data despite network partitions or server failures.

Because of the large size of AFS installations (numerous sites now have more than 2000 accounts), a number of system management tools have been developed. Backup software that understands that data can migrate between file servers and a load-monitoring system are two of the AFS system management components.

Miscellaneous

OTHER ISSUES: Remote program execution: The so-called *Butler System* comprising of a set of programs running on Andrew workstations. Its major goal is to make idle workstations available to users who might benefit from additional computing power (see [20,25] for Butler, and [12] for a survey on UNIX load-balancing schemes).

STATUS: AFS 3.0 has been a commercial product of Transarc Corporation since February 1990 available within the United States. It is expected to be available overseas in November 1990 due to a recent export ruling of the US Department of State. AFS 4.0 is presently being integrated with the other components of the OSF Distributed Computing Environment and is scheduled to be shipped by the OSF in mid-1991. AFS is running at about 40 sites.

CONTACT: Philip Lehman, Transarc Corporation, The Gulf Tower, Pittsburgh PA 15219, USA
e-mail: pll@transarc.com
REFERENCES: [18] – [36]

2.3 Cedar File System (CFS)

Main Goal

The Cedar File System (CFS) was developed as part of the Cedar experimental programming environment at the Xerox Palo Alto Research Center. CFS is a workstation file system that provides access to both a workstation's local disk and to remote file servers via a single hierarchical name space. Sharing of highly available and consistent file versions was the main goal of this project. CFS caches immutable copies of files from file servers on its local disk.

Advantages

The combination of the Cedar File System's semantics together with the tool for maintaining consistent versions of shared files is an interesting approach for a highly available and consistency-guaranteeing distributed file system. By making the files immutable, the cache on the workstation disk is guaranteed to have valid contents. Remote files may be given names in the local address space. These names are bound to a remote file by a remote file name and create date pair. This uniquely defines the file over all time. The remote file can be cached, and the cache can be used for many uses without intervention of the server.

The main disadvantage of this system is in file name resolution. Most resolutions are done locally with correct semantics, but some resolutions must access the file server. Operation of the system when a key file server is down is possible, but occasionally a needed resolution or a needed file cannot be done. A second disadvantage of this system is implied by the concept of immutability of files. Each version of a file is a complete copy resulting in an enormous amount of disk space and an increased disk allocation. There is no attempt made to reduce storage space by differentiating subsequent versions or by deleting unneeded versions. This has a small impact on the workstation, but has a large impact on the server. The goal of this work was to build a workstation file system, and to use existing file servers with as few changes as possible.

Description

The CFS is implemented by Cedar workstations (Dorado, Dolphin, and Dandelion workstations) and a collection of the IFS servers, XNS file servers, Alpine file servers, or NFS servers. (Sects. 8.36, 2.13, 2.1, 2.8, resp.). CFS accesses file servers through an internetwork using the File Transfer Protocol, XNS Filing Protocol, or the NFS protocol. The CFS is fully implemented inside the workstations code. CFS is designed to support sharing of file server remote files by workstations connected via a LAN. Each workstation is equipped with its own local disk. Diskless stations are not supported.

CFS provides each workstation with a hierarchical name space that includes the files on the local disk and on all file servers. Local files are private, remote files are sharable among all clients. Remote files may be given names in the local address space. To make a local file available to other clients, the file is transferred to a file server by giving it a remote name. Other workstation clients are now enabled to access this file by its remote name and to transfer it to their own local disk.

Remote file creation is atomic. Only entire files are transferable. It is now easy to understand why CFS only provides access to shared files which are immutable. An *immutable file* is a file that cannot be modified once it has been created. This leads to the naming strategy in CFS. Updating of shared files is accomplished by creating new file versions. For this reason, a complete file name consists of a server name, a root directory, some subdirectories (not necessarily), a simple name and a version. The server part identifies the server at which the file is allocated. Two variables provide access to the lowest (*!low*) and the highest (*!high*) versions of a given file. Even local files are treated by compilers and editors as immutable by always creating new file versions when results are written back to disk.

For security reasons, CFS uses the access control system provided by IFS for user identification. Access control lists contain user access rights. The information about a user – a user's name and his password – is passed to IFS as part of the FTP.

Availability plays an important role in CFS and is achieved by replication. Important directories are automatically as well as periodically replicated to multiple servers by a daemon process. When a remote file is opened for read and it is not already cached, then it is transferred to the workstation cache. All read requests to open remote files are satisfied from a local cache. For this purpose, every workstation uses a portion of local disk space as cache. The cache is managed using an approximate LRU (least recently used) strategy.

Consistency of files is achieved by the concept of immutability of files – creating a new file version every time a particular file is modified. Consistency of file server directories is achieved by performing what is logically a transaction for directory updates. Updates are indivisible and serialized. Thus, transferring a new file to a server, assigning it a new version number, and entering its name in the file server directory appear to be a single act.

Miscellaneous

OTHER ISSUES: An application of the Cedar File System is the support of the so-called *Cedar Software Management Tools* known as the *DF system*. DF provides an environment for the handling and maintaining of large software packages. A programmer using some DF files can be sure to have a consistent view over these subsystem files.

Cedar is also a programming language. The Alpine File System has been written in the Cedar language.

STATUS: Research on CFS is active and continues to be enhanced. CFS is still in use as an integral part of the Cedar Programming Environment.

CONTACT: David K. Gifford, Lab. for Computer Science, MIT, 545 Technology Square, Cambridge MA 02139, USA, or Mark Weiser, Xerox Palo Alto Research Center, 3333 Coyote Hill Road, Palo Alto CA 94304, USA

e-mail: gifford@brokaw.lcs.mit.edu or weiser.pa@xerox.com

REFERENCES: [37] – [45]

2.4   Coda

Main Goal

Coda is a highly available distributed file system for a large collection of portable and non-portable UNIX workstations. Coda is location, access, concurrency and failure transparent and provides optimistic replication transparency.

Advantages

Coda provides high availability without sacrificing the many merits of its ancestor, AFS-2 (cf. Andrew, Sect. 2.2). The design of Coda optimizes for availability and performance, and strives to provide the highest degree of consistency attainable in the light of these objectives. Usage experience and measurements from the prototype show that the performance cost of providing high availability in Coda is indeed reasonable.

Description

Coda is available today for IBM RTs, DEC Pmax's, and Toshiba I386 laptops. UNIX applications run on Coda without recompilation or relinking.

Resiliency to server and network failures is attained through the use of two distinct but complementary mechanisms. One mechanism, *server replication*, stores copies of a file at multiple servers. The other mechanism, *disconnected operation*, is a mode of execution in which a caching site temporarily assumes the role of a replication site. Disconnected operation is particularly valuable when using portable workstations. From a user's perspective, transitions between connected and disconnected modes of operation are seamless.

When network partitions occur, Coda allows data to be updated in each partition but detects and confines conflicts. Most directory conflicts are transparently resolved by the system. A repair tool is provided to assist users in recovering from all other conflicts. This approach to conflict resolution is an adaptation of that originally proposed by LOCUS (Sect. 4.12).

Each server operation in Coda typically involves multiple servers. To reduce latency and network load, Coda communicates with replication sites in parallel using the MultiRPC facility of the RPC2 remote procedure call mechanism. MultiRPC has been extended to use hardware multicast support, if available.

Coda uses token-based authentication and end-to-end encryption integrated with RPC2.

Coda servers use the Camelot transaction facility (Sect. 8.11) for atomicity and permanence of meta-data. Neither nested nor distributed transactions are used. Camelot is in the process of being replaced by RVM, a lightweight recoverable virtual memory facility.

Miscellaneous

Almost all the server and client code for Coda is implemented as user-level processes. The one exception is a small amount of code, called the *MiniCache*, that resides in the client kernel as a VFS driver for the SUN-Vnode interface.

STATUS: The Coda prototype is in daily use by a small user community at the School of Computer Science in Carnegie Mellon University. Work is under way to complete and tune the implementation, and to support a larger user community.

CONTACT: Mahadev Satyanarayanan, School of Computer Science, Carnegie Mellon Univ., Pittsburgh PA 15213, USA

e-mail: satya@cs.cmu.edu

REFERENCES: [33], [46] – [48]

2.5 Extended File System (EFS)

Main goal

The main motivation behind the MASSCOMP (Massachusetts Computer Corporation) Extended File System for UNIX approach (EFS) was the desire to bring full transparent access to files on remote computer to all users. The goal of transparency has been achieved by building the concept of remoteness into the existing UNIX file system I/O architecture.

Advantages

EFS is a system that operates in a multiprocessor environment. In its 1985 form it provides transparent remote access only to disk files. Remote devices and diskless nodes are not supported. The concept of process pools, i.e., available kernel processes, which can play agent roles for any EFS client request, is the main advantage of EFS. No single client process is able to block an EFS server from serving other client processes.

Description

EFS is client/server based and operates between two or more MASSCOMP computers on the same Ethernet, all running an enhanced version of the Real-Time UNIX (RTU) kernel, a MASSCOMP variant of the UNIX time sharing system. A new *reliable datagram protocol* (RDP) has been designed and implemented providing guaranteed message delivery and the ability to dynamically create multiple low-overhead connections through a single socket. The IPC uses the UNIX 4.x BSD socket mechanism.

Because EFS works in an UNIX environment, an inode contains all information with regard to a local file. A *rinode* is a special variant of an inode. It contains a machine-Id which uniquely identifies a single host on the network as well as the address of an usual in-core inode within the memory of that host. Whenever a remote file is involved, the rinode information is used. Mounting enables network-wide access to remote files.

Protection across the EFS is built in as follows: All the machines in the network are converted to the same password file, so that nearly the normal UNIX protection mechanisms can be used with the exception of selective mount points. These are added to make certain files on some machines inaccessible to non-authorized users. The global protection domain is used to simplify the concept of file ownership.

The client/server model in EFS is stateful. All remote file operations are based on transactions. The client process is blocked until the server completes the transaction.

Miscellaneous

STATUS: Research should be going on to eventually support remote devices and diskless nodes. The current status is unknown.
REFERENCES: [49]

2.6 HARKYS

Main Goal

The main goal of the HARKYS project was to provide a distributed file system both location transparent and reliable for different machines running different UNIX systems without modification to the kernel.

Advantages

HARKYS is easily portable and makes the joining of UNIX systems as different as 4.x BSD and System V in one global file system possible. However, locking problems will have to be solved subsequently.

Description

HARKYS can only be installed above an existing UNIX system. Programs using remote files must be relinked with a new C-library, but without any change to the sources. Compatibility with 4.x BSD, the System V family, AIX and ULTRIX is given. HARKYS is presently running on various machines, e.g., VAX and SUN.

RPC is used for client/server communication based on standard protocols. Up to now TCP/IP is supported. A software layer is placed between the user and the operating system interface which forwards remote requests to the corresponding server processes.

These are created by a daemon, called *spawner*, for each participating operating system separately. In addition, the server process makes any data conversions required.

Remote hosts need to be mounted. If a path name contains a mount point, the translation of a new path name starts from the remote root directory. This way, the remote mount point has to be the root.

For the purpose of access control, the standard UNIX mechanism is used although more general access strategies can also be employed. Rights to other hosts are described in an authentication database which is included at each host.

The client process maintains all information concerning remote files, some information being replicated at the server's side. In the case of a crash of a process or a host or in the case of network problems, all remote and local resources are released with the help of this data, and participating processes are terminated.

Miscellaneous

STATUS: The system has been tested on VAX/750 and SUN systems. Future work includes the extension of accounting mechanisms across the network, the utilization of other transport protocols, and the provision of system administrator network tools. The current status is unknown.

CONTACT: A. Baldi, Research and Development Division, Systems and Management S.p.A., Pisa, Italy

REFERENCES: [50]

2.7   IBIS

Main Goal

The purpose of IBIS is to provide a uniform, UNIX compatible file system that spans all nodes in a network. IBIS is being built as part of the TILDE project (Transparent Integrated Local and Distributed Environment, Sect. 8.66) whose goal it is to integrate a cluster of machines into a single, large computing engine by hiding the network. IBIS, a successor of STORK (Sect. 8.62), provides three levels of transparency: file access, replication and concurrency.

Advantages

IBIS exploits location transparency by replicating files and their migration to the place where they are needed. Replication and migration of files improve file system efficiency and fault tolerance.

The main disadvantage of IBIS is quite an anomaly. Since IBIS is implemented as a separate layer between UNIX kernel and the applications, for local file accesses, the overhead rises with the size of the file resulting in a poor performance. Placing IBIS into the kernel should speed it up.

Description

IBIS runs on VAX-11/780 hosts under the operating system UNIX 4.2 BSD connected via a 10 Mbit/second Pronet. Because IBIS implements all remote file operations with the IPC facility of UNIX 4.2 BSD and is based on the standard TCP/IP protocol, it is not restricted to LANs. Even if a remote host running TCP is connected via gateways or lossy subnets, it can participate in IBIS.

IBIS provides UNIX-like file names (path name strings) in a hierarchical name space. Such a path name consists of *hostname*, i.e. no location tranparency, and a *local name*. The local name is a standard UNIX path. IBIS provides access transparency since the same access operations can be used for both local and remote files.

The access transparency layer is sandwiched between UNIX kernel and its applications. For remote files, this layer uses a RPC protocol for interacting with the server. Not the whole UNIX system call semantics is implemented by IBIS. Some UNIX commands are missing, some are replaced by look-like UNIX commands. For example, *rm* does not remove a file. Instead, it moves the file to a special directory called *tomb*, from where it will eventually be removed. With symbolic links crossing machines, a user can construct a directory tree spanning several machines.

Security issues in IBIS are implemented via authentication. Both sites, client as well as server, need a way of ascertaining the authenticity of each other. IBIS solves this problem as follows: A dedicated server is established by an authentication scheme called a *two-way handshake via secure channel*. A client reserves a port number and executes a so-called *connection starter* program. This starter opens a secure channel to the so-called *server creator* in the remote host and passes the user-Id to it. The server creator process establishes a server with access rights, as indicated by the user-Id, and reserves a port number as well. Exchanging port numbers via the secure channel from now on allows the establishment of a connection verifying each other's portnumber. The secure channel is implemented by making both, connection starter and server creator, privileged processes which communicate via privileged port numbers.

Availability in IBIS is achieved by replication. There are two main differences with regard to LOCUS (Sect. 4.12). IBIS avoids both the bottleneck of a synchronization site (LOCUS's CSS) as well as remote synchronization for local replicas. To achieve this, IBIS uses a decentralized control scheme. A primary copy strategy is used with the following update protocol. If the primary copy is updated, the update broadcasts a signal that invalidates all cached copies. The use of strict state transition rules provides consistency. The IBIS replication scheme, which was later adopted by Andrew (Sect. 2.2), is called *Staling*, while the Andrew people called it *Call-back*. Andrew originally used polling on every file open(for read), which is a "poor" choice since reads are more frequent than writes. Staling is preferable if reads are more frequent than writes.

Miscellaneous

OTHER ISSUES: Migration is used to relocate files so as to speed up write requests (demand/forced replication). IBIS provides explicit commands for migration. Moreover, consistency can be sacrificed for a simplification of the update protocol.

STATUS: Research is going on to develop automatic migration and placement of files.
CONTACT: Walter F. Tichy, University of Karlsruhe, Informatik, Postfach 6980, D-7500
Karlsruhe, Germany
e-mail: tichy@ira.uka.de
REFERENCES: [51], [52]

2.8 Network File System (NFS)

Main Goal

The SUN Network Filesystem (NFS) was implemented to solve the problem of communication and sharing in a heterogeneous network of machines and operating systems by
providing a shared file system access- and location-transparently throughout the network.

Advantages

NFS is designed such that it can easily be transferred to other operating systems and
machine architectures. It therefore uses an *External Data Representation* (XDR) specification. It describes protocols independently of machines and systems. In addition,
the source code for user-level implementation of the RPC and XDR libraries has been
published. Most importantly, NFS has become a de facto standard.

The main disadvantages of NFS are that it does not support all of the UNIX file
system semantics. For example, due to a stateless protocol, removing open files or file
locking are not supported. There are no mechanisms for concurrency control.

Description

NFS runs on various machine types and operating systems: UNIX 4.x BSD, SunOS,
DEC Ultrix, System V.x, VMS and MS/DOS etc.

Historically, NFS is part of the SunOS operating system implemented as a software
system for accessing remote files across LANs. It is implemented on top of a Remote
Procedure Call package in a client/server manner and adds a new interface to the kernel.

It is possible for any NFS-machine to be a client, a server or both. The client side is
implemented inside the UNIX kernel. The mount service is implemented as an everlasting
daemon process which is started every time a mount request is initialized.

The file system interface consists of two parts, first, the *Virtual File System* (VFS)
interface which defines operations on the entire file system and second, the virtual node
(*vnode*) interface which defines the operations on a single file within the file system.

Communication is synchronous and uses a stateless protocol (in comparison, for example, RFS (Sect. 2.9) that uses a stateful protocol). Because a server does not need to
keep track of any past events, crash recovery is easy to implement. NFS uses the DARPA
User Datagram Protocol (UDP) and the Internet Protocol (IP) for its transport level.

NFS uses normal cache-buffers (read-ahead, delayed-write; write-on-close in an earlier
version) to increase performance. For read-requests, performance is good enough to

neglect local or remote access differences, however performance for write-requests is poorer because it is synchronous at the server site.

Naming is done by using UNIX path names. The NFS servers export complete file systems. Clients can mount/attach any file system branch of an exported remote file system on top of a local directory (mount cascades are provided). Thus a hierarchical, uniform name space is presented to the users individually. Any locally attached file system can be named location transparently. Path name traversal and lookups are performed remotely at the server sites. Crossing mount points results in a rather expensive scheme since every new lookup needs another RPC to the server involved. Therefore, caching is used to increase the lookup performance.

The RPC package allows an encrypted authentication method, while still allowing access with a simpler authentication method for PC-users. However, due to the distributed character of a network file system, the security features provided are weaker than those of standard UNIX. NFS uses UNIX-like permission checks between client and server sites. User-Id and group-Id are passed within the RPC and are used by the server to decide whether the client request for access can be satisfied or not. A server can easily protect its local file system by an export list of exported file systems and the authorized clients that can import each of them.

A server does not do any crash recovery at all. When a client crashes, no recovery is necessary (due to the stateless protocol). If a server crashes the client resends requests until a response is received. Typically, NFS requests are idempotent. This stateless server implementation implies some UNIX incompatibilities. For example, if one client opens a file and another removes it, the first client's reads will fail even though the file is still open.

A useful extension to NFS is the so-called *Yellow Pages* (YP) mechanism, i.e. a service that provides distribution and central administration of read-only system databases. The exported file system information is maintained on each machine and made highly available this way.

Miscellaneous

Extensions or enhancements to the pure NFS implementations are currently undertaken, for example at Cornell University (RNFS, Sect. 8.53) and at DEC WRL (Spritely NFS, Sect. 2.11).

OTHER ISSUES: NFS Version 4.x supports diskless workstations; mounting of their own root directory from servers is provided.

STATUS: NFS is an established product and support on 4.x BSD is available through SUN and Mt. Xinu, on System V through Lachman Associates and Unisoft, and on Ultrix through DEC. The VMS implementation is for the server side only. The MS/DOS part is complicated by the need for a redirector layer that redirects system calls to the appropriate MS/DOS services.

CONTACT: SUN Microsystems, Inc., 2550 Garcia Ave., Mountain View CA 94110, USA

REFERENCES: [53] – [67]

2.9   Remote File Sharing (RFS)

Main goal

Remote File Sharing (RFS) is one of the networking-based features offered in the AT&T
UNIX System V, providing location-transparent access to remote files and devices.

Advantages

RFS supports 100% of the UNIX file system's semantics. This means that, in contrast
to NFS (Sect. 2.8), file locking and append write are possible. Thus, existing user
applications can use the features of RFS without modification or recompilation.

The main disadvantages of RFS are first, that only one of the UNIX versions (UNIX
System V) is supported, and second, that a new special protocol – not RPC-based –
makes it hardly portable.

Description

RFS runs on multi-vendor hardware operating under UNIX System V connected through
an Ethernet or a Starlan LAN. RFS is based on a client/server model. Every machine can
be a client, a server or both. RFS is a kernel extension that deals with file access requests
as follows: The mount-service enables server machines to make remote files available
to client machines. The RFS interface decides whether a request is local and can be
satisfied at the local UNIX file system, or remote. If a request is remote, communication
is established via the Streams I/O system, developed for the UNIX time-sharing system
for intermachine communication.

RFS has implemented a remote mount service allowing a user to add remote file
system branches to its local file system tree. Subsequently, remote requests to these
files are handled as if they resided inside the local machine. The notion of *domains* –
an administrative name space made up of a collection of machines – was designed by
RFS. For each domain, a name server exists. Name resolution is done by using this RFS
name server which is modeled as a transaction handler: it receives requests, performs
the operation desired, generates and sends back a reply to the originator. A so-called
advertise command gives a user the possibility to make subtrees of its local file system
mountable for clients.

Communication between name servers on different machines is done by the standard
Transport Layer Interface (TLI). The TLI is based on the ISO Transport layer and is
designed to support higher layer protocols in a protocol independent way. Both virtual circuit and datagram services are provided, enabling the use of different standard
transport protocols (TCP/IP, ISO, XNS, SNA).

Concurrency control is achieved by file locking. The RFS uses a stateful protocol.
Therefore, the server has to maintain information about the current state of all of its
clients. Because it would be difficult and costly for the client to rebuild the server's state
after a server crash, no server crash recovery is done.

Mapping of remote user-Id or group-Id to local user identifications is provided. This means, for example, that remote superusers can be mapped onto a local standard restricted user-Id.

Miscellaneous

RFS allows access to remote special devices so that expensive peripheral devices can be shared economically.

STATUS: RFS is an integrated part of the commercial UNIX System V developed by AT&T.

CONTACT: AT&T, 190 River Road, Summit NJ 07901, USA

REFERENCES: [68] – [73]

2.10   S-/F-UNIX

Main Goal

The main goal of this project at the University of Illinois in the early 1980s was to extend the widely known and used UNIX time-sharing operating system to an access transparent distributed operating system. Performance of the communication software has been the major focus of research, while keeping most of the UNIX semantics.

Advantages

The project has shown that communication based on virtual circuits can achieve high transfer rates, fast response, and low local overhead costs. A disadvantage of S-/F-UNIX is its restriction to 64kB in regard to path names, read or written data. Terminal-to-terminal communication and interprocess pipes are not supported.

Description

S-/F-UNIX is a client/server system. *S-UNIX* is the specialized operating subsystem that runs user processes. The subsystem running on the file server is called *F-UNIX*. DEC PDP-11 line hosts communicate with each other through a high-bandwidth virtual circuit switch. Small front-end processors handle the data and control protocol for error and flow-controlled virtual circuits. The conventional UNIX kernel was modified. Nodes are connected via the Datakit switch, which is a switch that combines the desirable properties of both packet and virtual circuit switching while still providing high and dynamically allocatable bandwidth, since it uses packet switching with demand multiplexing in its internal implementation. A so-called (*NK network kernel*) protocol has been implemented to speed up transmission.

File space has been extended by mounting an entire disk volume on top of an existing directory. S-UNIX allows its user to mount a F-UNIX file server in an analogous way. Multiple mounts of both kinds can be active simultaneously. File names are hierarchical (path name strings). S-/F-UNIX uses remote inodes (*rinodes*) every time a remote file is accessed. It contains all the information needed for the S-UNIX subsystem to handle a specified file. This implies a pointer identifying the related F-UNIX machine holding the file and a unique number assigned by that F-UNIX machine. All other information about the file is maintained at the remote site.

Assuming that the operating system and the communication mechanisms are trustworthy, S-/F-UNIX avoids having to deal with problems of authentication beyond those present in the current UNIX system.

Miscellaneous

OTHER ISSUES: Special files (devices) are treated like special files in UNIX. They are handled by dedicated F-UNIX subsystems.
STATUS: Not active.
CONTACT: G.W.R. Luderer, AT&T Bell Laboratories, Murray Hill, NJ 07974, USA
REFERENCES: [74] – [77]

2.11 Spritely NFS

Main Goal

Spritely NFS is an experiment to isolate the performance effects of using an explicit cache-consistency protocol.

Spritely NFS is a distributed file system (just like NFS, Sect. 2.8), and provides the same levels of transparency, except that it provides correct concurrent access from multiple clients which NFS does not do.

Advantages

Spritely NFS showed that removing the "stateless server" restriction of NFS significantly improves performance overall, and does not cause more than a small degradation in performance in any cases.

A crash-recovery protocol has not been implemented. This has been shown to be possible by the Sprite project (Sect. 4.20), but it would have required a lot more code within the current Spritely NFS implementation.

Description

Currently Spritely NFS runs, as a new-NFS-kernel-implementation, on proprietary hardware, but the code should be portable.

Spritely NFS makes use of XDR, SUN-RPC, and UDP/IP.

Concurrency is controlled by an explicit cache-consistency protocol, which allows clients to cache files if and only if there are no writers, or if there is one writer and no other active client.

Consistency state must be recovered after a server or client crash. Spritely NFS does not implement this; Sprite does, or the "Leases" mechanism might be used. For more details, see the NFS section (Sect. 2.8).

Miscellaneous

STATUS: Spritely NFS is a research prototype, not available for others, and is not useful for a production environment.
CONTACT: Jeffrey Mogul, Digital Equipment Corporation Western Research Laboratory, 100 Hamilton Avenue, Palo Alto CA 94301, USA
e-mail: mogul@decwrl.dec.com
REFERENCES: [78], [79]

2.12   VAXcluster

Main Goal

The Digital Equipment Corporation developed the VAXcluster product as a set of closely-coupled DEC VAX computers providing a highly available and extensible configuration with access, concurrency, and failure transparency. Another key concern was the performance of the system.

Advantages

The implementation of a high-speed message-oriented computer interconnect has increased the system's performance. In contrast to other highly available systems, VAXcluster is built from general-purpose, off-the-shelf processors, and a general-purpose operating system.

The main disadvantage of VAXcluster is that it does not provide location transparency.

Description

Each VAXcluster host runs a distributed version of the VAX/VMS operating system and is connected to the *CI* (Computer Interconnect) through a *CI port*. CI ports are device-specific and have been implemented so far for the VAX 11/750, 780, 782, 785, and VAX/8600 hosts. The CI logically is a bus. Physically, it is a dual path serial connection with each path supporting a 70 Mbit/second transfer rate.

The *SCA* (System Communication Architecture) is a software layer that provides datagram-style and VC communication. To ensure delivery of messages without duplication or loss, each CI port maintains a VC with every other cluster port. RPC-based communication is supported too.

The *Hierarchical Storage Controller* interprets a so-called *Mass Storage Control Protocol* (MSCP) which separates the flow of control and status information from the flow of information to increase performance.

VAXcluster provides a cluster-wide shared file system to its users. A complete file name includes the disk device name (a system-wide unique name), the directory name, and the user-defined name itself. The disk device name gives the information concerning which node a particular file is located on. Therefore, location transparency is not achieved.

Password information resides in a single file shared by all VAXcluster nodes. A user obtains the same environment regardless of the node he is logged into.

A *lock manager* is the foundation of all resource sharing. It provides services for naming and locking of cluster-wide resources. A cluster connection manager is responsible for coordination within the cluster. It recognizes recoverable failures in remote nodes and provides data transfer that can handle such failures transparently.

To prevent partitioning, VAXcluster uses a quorum voting scheme (see [14] for a survey on voting approaches). Initial vote and quorum values are set for each node by a system administrator.

Miscellaneous

STATUS AND CONTACT: VAXcluster is a product still supported by several engineering groups at Digital Equipment Corporation. For the early design concepts, contact Hank Levy, Dept. of Computer Science and Engineering, FR-35, Univ. of Washington, Seattle WA 98195, USA
e-mail: levy@cs.washington.edu
REFERENCES: [80]

2.13   Xerox Distributed File System (XDFS)

Main Goal

The Xerox Distributed File System (XDFS) from Xerox PARC was a research project to design a very robust multiple-server system. It was intended to provide a basis for database research. Besides, XDFS is location, access, and concurrency transparent.

Advantages

XDFS was an early approach in designing a distributed file system with sophisticated features such as stable storage, transaction handling, and concurrency control mechanisms. The main disadvantages of this system are the great overhead involved in the implementation of stable storage and the limited amount – a page – of data to be transferred as part of a single request.

Description

XDFS written in MESA language, runs on an Alto minicomputer, and communicates using an Ethernet communication system. Client communication in XDFS is built on the *Pup* (PARC Universal Packet) level, which implements an internetwork datagram.

The XDFS uses a B-tree to map file-Id and page number to a disk address. Thereby, access and location transparency are provided. It provides fine-grained locking at the byte level of a file. Access control issues are simply based on the user-Id. This limits the degree of security provided.

For concurrency control (single writer, multiple reader), XDFS uses time-limited breakable locks. Deadlocks are detected through timeouts. All operations are repeatable. A two-phase commit protocol ensures correct updates.

Failure handling and recovery are supported by shadow pages plus intention logs (redo logs) which are stored at atomic stable storage. In XDFS, a client needs to set up a transaction before accessing or creating any file. All committed updates are completed immediately after a crash. If a new crash appears while the recovery procedure is active, the whole process will be restarted.

Replication is not supported.

Miscellaneous

OTHER ISSUES: XDFS supports the database management system of Cedar (Sect. 2.3).
STATUS: The first version of XDFS became operational in 1977. XDFS was shut down in February 1983 and replaced by the Alpine project (Sect. 2.1).
CONTACT: Bob Hagmann, Xerox Palo Alto Research Center, 3333 Coyote Hill Road, Palo Alto CA 94304, USA, or J. Dion, Computer Laboratory, Cambridge University, Corn Exchange Street, Cambridge CB2 3QG, UK
e-mail: hagmann.pa@xerox.com
REFERENCES: [10], [81]

Chapter 3

Object-Oriented Distributed File Systems

3.1 DOMAIN

Main Goal

The DOMAIN system is mainly a distributed file system where an object-oriented approach is taken. It is a commercial product of Apollo Computers, Inc., that connects personal workstations and server computers. All processors transparently share a common network-wide virtual memory system that allows groups of users to share programs, files and peripherals.

Advantages

DOMAIN provides an efficient network-wide single-level store mechanism. However, in some cases an IPC mechanism would be preferable. Not much work has been done to make the system more reliable.

Description

DOMAIN is an object-oriented distributed file system for Apollo Computers. The network is a 12 Mbit/second token ring. Connections to other networks can be integrated. The DOMAIN user environment offers AUX, a UNIX System III compatible software-environment with 4.2 BSD enhancements. DOMAIN and AUX programs can be mixed freely.

The only way to communicate is via shared memory supported by the single-level store mechanism. The object-based virtual memory allows pages to be transferred across the network. To allow recovery from link failures, the network can be installed in a star-shaped configuration.

Files are objects and may be of different types. It is possible to implement file operations tailored to specific file types. Each object has a unique identifier (UID) which is generated by concatenating the unique node-Id of the node generating the object with a timestamp from the node's timer. The objects are located by a sophisticated search algorithm which uses hints for the most likely location. The permanent storage of an object is always entirely on one node, most often the original one. All operations on an object are performed on this node. Objects have a type UID that identifies a *type manager* handling this object. Users are provided with hierarchical names that are translated by local name servers into the object's UID. The network-wide *root* directory is replicated on every node.

Users are authenticated by the use of a replicated system-wide user registry. A user identifier consists of a person, project and organization identifier and the login node. Each file system object is associated with an *access control list* object containing a UNIX-like list of rights.

The system detects concurrency violations by the use of version numbers, but does not provide transparent serialization. The distributed lock manager offers the possibility of either *many readers* or *single writer* locks or a *co-writers* lock – which allows multiple writers that have to be co-located to work at a single network node. In addition, this manager is responsible for the management of the caches. A write-through approach is used for updates. Validation of cached data is initiated by the clients.

Miscellaneous

OTHER ISSUES: A generic service tool provides among other things a calendar utility, a mail program and a document editor.

Diskless nodes can be incorporated.

STATUS: The Apollo DOMAIN system has been in full commercial production since the middle of 1981. Thousands of networks and over 20 000 workstations have been installed. The largest single network consists of more than 1800 nodes.

CONTACT: Paul H. Levine or Paul Leach, Apollo Computer, Inc., Chelmsford MA 01824, USA

e-mail: pjl@apollo.hp.com

REFERENCES: [82] – [92]

3.2   Helix

Main Goal

Helix is an object-oriented distributed file system for the XMS system developed at BNR in Ottawa. Its main objectives are the support of a diversity of applications, capability-based security, and the provision of access, location and failure transparency.

Advantages

Helix has some useful features for failure resistance and protection of files. The server CPU could become a bottleneck since the server has to do most of the work (of path name translation, checking access rights, etc.). The system seems to be configured solely for the software development at BNR, since it is hardly portable to other systems.

Description

The XMS system is being run on Motorola M 68010-based workstations. Helix is compatible with the UCSD file system.

Helix uses the XMS blocking communication primitive, called *rendezvous*, but nothing – as far as the author knows – has been published about the underlying protocol and the semantics of a rendezvous. The communication between multiple LANs is done by special communication servers and is based on the ISO OSI model. Helix is extended to multiple LANs by a server called network-Helix which implements the standard interface and passes requests to the far end via virtual circuits.

Each object has a unique capability containing the server-Id, a field of rights, the number of the object and some random bits to make it unforgettable. Capabilities are stored in directories together with a user-defined string. Directories can be linked freely, not necessarily hierarchical. The naming of objects can be done either directly by a capability or by a path name. Requests for remote files are passed from a so-called *client Helix* to the remote server that does path name translations and capability checks. Files are grouped in logical volumes, typical equivalent to file servers. Links between different volumes are not permitted. Caching is used to improve the performance.

The capabilities which are created by the system are automatically encrypted and decrypted.

The system allows transactions, which can be used for updates, and the creation of instances for a file. Transactions can also be applied on sets of files, or otherwise can be nested. Some measures are taken to improve reliability: Unreadable loops which are possible in a non-hierarchical name space are discovered and linked into a special directory for manual resolution. In case of program failures, all files are closed or aborted. Clients are periodically checked to detect crashes.

Miscellaneous

STATUS: The system was in use at BNR with 15 LANs and close to 50 file servers which were supporting almost 1000 workstations. The current status is unknown.
CONTACT: Marek Fridrich, Bell-Northern Research, PO Box 7277, Mountain View CA 94039, USA
REFERENCES: [93] – [95]

3.3 SWALLOW

Main Goal

SWALLOW was a distributed file system supporting highly reliable object-oriented data storage. It provided all kind of transparency levels, i.e., those of location, access, concurrency, failure, and replication.

Advantages

SWALLOW was yet another object-oriented DFS. Each update created a new version of the object. However, only the updated pages were copied. All other pages were merely taken over by the new version. It could manage objects of arbitrary size and provided synchronization and recovery mechanisms as part of the object model rather than on top of it.

Its main disadvantage was the handling of mass storage. Since every update operation created a new version, a large amount of data needed to be stored. Performance has never been evaluated.

Description

SWALLOW ran on standard UNIX hardware in a client/server behavior. A so-called *broker* software package provided access as well as location transparency for the client machines. Therefore, each client machine needed to run a broker copy.

SWALLOW was a distributed file system based on an object model. All data entities were encapsulated in objects. Its files were stored in remote, highly autonomous servers called *repositories*. The core of a repository was a stable append-only storage called *Version Storage (VS)*. VS contained the version history of all data objects as well as information needed for crash recovery. Every time an object was modified, a new version of the particular object was appended to the VS. Updates were embedded in transactions. In case of failures, object histories provided a backward recovery mechanism. Updates which were not yet committed, were stored on stable storage.

It used an asynchronous datagram communication. Responses to requests were collected in an asynchronous fashion by the broker. A special end-to-end protocol (SWALLOW Message Protocol SMP) based on datagrams was implemented for data transfer.

A capability-based protection scheme was implemented. Concurrency control – single writer, multiple reader – was achieved by timestamps. Each object was protected by a monitor providing mutual access control. The monitor state itself was kept in volatile memory and thus, after a crash, all objects were automatically unlocked.

Data reliability was provided in a simple replication scheme. A duplicate was stored by a second storage device, and an update was applied immediately to both devices.

Miscellaneous

STATUS: SWALLOW was a data storage system designed in early 1980 at the Laboratory for Computer Science of the Massachusetts Institute of Technology. The main design components, repositories and brokers were prototypically implemented and tested under artificial workload. Not active.
CONTACT: Liba Svobodova, IBM Zürich Research Lab., Säumerstr. 4, CH-8803 Rüschlikon, Switzerland
e-mail: svo@ibm.com
REFERENCES: [96] – [98]

Chapter 4

Traditional Distributed Operating Systems

4.1 Accent

Main Goal

Accent is a successor to the RIG system (Sect. 8.54), a message-based network access machine. It tries to solve the problems of transparent access, failure notification, protection and access to large objects (for example files) with low message overhead. Also, it provides location and failure transparency.

Advantages

Accent achieves these goals efficiently by introducing capabilities for message ports and by the use of copy-on-write virtual memory management throughout the network.

The disadvantages of the system are its restriction to one type of hardware and that UNIX software cannot be run on it.

Description

Accent consists of a new operating system kernel which operates only on PERQ workstations. The IPC facility is supported under a modified version of UNIX 4.2 BSD.

The inter-process communication of Accent uses ports as an abstraction to which messages can be sent asynchronously. Multicast is not provided.

Each process can have various ports, for example a kernel port. The rights for possessing a port and for sending or receiving messages are distinguished. The semantics are application dependent, but at-most-once semantics can be achieved. If a port is full, the sending process will be suspended or will be informed of this. Rather than copying large messages all at once, they are handled copy-on-reference even across the network, making virtual memory possible. All communication over the network is done by special network servers running on each host which forward messages transparent to the senders. Each network server maintains a mapping between network ports (ports which are remotely accessible) and corresponding local ports. The use of these servers has the advantage that they can do data type conversions as well as the implementation of secure transmission outside the kernel.

Every access to a port is protected by the use of capabilities, given that every kernel is trusted. The network servers map local to remote capabilities.

Processes are allowed to move from one machine to another. Process migration is transparent to other processes but all network servers must get the information.

Since ports differentiate between the owning and receiving process, it is possible to give the surviving process of a crash both rights automatically to assure further computation. Additional reliability issues must be implemented on top of the given system.

Miscellaneous

STATUS: Accent is completed and used on 150 PERQ workstations at Carnegie Mellon University. It leads to the evolution of MACH (Sect. 4.13).
CONTACT: Richard F. Rashid, Computer Science Dept., Carnegie Mellon Univ., Pittsburgh PA 15213, USA
e-mail: rashid@cs.cmu.edu
REFERENCES: [99] – [104]

4.2   Cambridge Distributed Computing System (CDCS)

Main Goal

The main intention at Cambridge was the use of the Cambridge Digital Communication Ring by a coherent system employing the "pool of processors". The Cambridge Distributed Computing System (CDCS) was primarily composed of a processor bank and some servers. It provided access, location and failure transparency.

Another aim was the sharing of common services (such as printing, disc storage, and electronic mail) in a heterogeneous environment.

Advantages

The CDCS was one of the first heterogeneous distributed system approaches. Despite this fact it provided good performance and reliable services and was used as the research environment at the Computer Laboratory for ten years. It showed that processor bank management and the support for service invocation by heterogeneous subsystems form a reliable basis for distributed operating system design.

The main disadvantage in the early days was that CDCS was hardly portable, since it was tailored to very specific hardware, the Cambridge Ring.

Description

CDCS was built as a client/server model using a processor bank. The processor bank consisted of LSI-4 and Motorola 68000-based machines. Most of the servers were running on Z-80-based computers or on a PDP 11/45. It was augmented by VAX UNIX systems and a number of Ethernet-based MicroVax2s for large-scale program development. Some interworking with SUN and XEROX Ethernet-based distributed systems was also achieved. Any operating system could be loaded into processor bank machines, including experimental systems under development. The widely used systems were the TRIPOS single-user operating system and the Mayflower kernel.

Initially, CDCS was implemented on a single Cambridge Ring. Later, it was extended to operate over three bridged Cambridge Rings, and even by 1 Mbit/second satellite broadcast channel over a wide area. This was done in the *Universe* project. Universe stands for Universities Extended Ring Satellite Experiments.

A protocol hierarchy was developed: the Basic Block Protocol (BBP) above the Cambridge Ring, a Single Shot (request/response or transaction) Protocol (SSP), and a Byte-Stream Protocol (BSP) at the next layer. Specialized application protocols were implemented above these basic protocols.

In the *Mayflower* project, set up in 1982, CDCS got its concurrent programming language, Concurrent CLU (CCLU). This language included a language-level RPC facility. The RPC ran over UDP/IP on the Ethernet and over BBP on the Cambridge Ring. It operated through bridges and across a Ring-Ethernet gateway. The CCLU RPC semantics were at-most-once, but optionally exactly-once semantics could be offered, provided there were neither server crashes nor prolonged network failures.

Flat text names were used for naming. The names were mapped to ring addresses and port numbers. The mapping was done by a name server residing at a well known address within the network. The results of a name server lookup were cached and used as hints for subsequent invocations. If the name was found, the name server returned the machine number, the port and the protocol it expected. Files of the user's TRIPOS file system were automatically mapped by stubs to files of the file server.

Each processor bank operating system used its own authentication scheme. However, it registered its user with the CDCS authorization server, the so-called Active Object Table (AOT). AOT provided a session key, which together with user information, functioned as a capability for use of system services.

The file server distinguished normal and special files which were atomically updated. A garbage collector program checked the file system for inaccessible files which were then deleted. Availability of the system was improved by the use of a processor bank. The main approach for fault tolerance was to bring up servers after a crash by detecting the crash and downloading the appropriate programs.

Miscellaneous

STATUS: CDCS was designed in the early 1980s. It was in everyday use at the Computer Laboratory at Cambridge until December 1989.

A new lightweight kernel WANDA (Sect. 4.22), capable of running on VAXs, Firefly multiprocessors, and M68000-based systems, has been developed. The new distributed systems research environment is based on the Cambridge Fast Ring.

CONTACT: Roger M. Needham, Computer Laboratory, University of Cambridge, Corn Exchange Street, Cambridge CB2 3QG, UK

e-mail: rmn@computer-lab.cambridge.ac.uk

REFERENCES: [105] – [112]

4.3 Charlotte

Main Goal

Charlotte is a distributed operating system developed and in production use at the Computer Sciences Department of the University of Wisconsin–Madison. It provides location and access transparency. It is part of the Crystal project that was funded starting in 1981 to construct a multicomputer with a large number of substantial processing nodes.

Advantages

The Charlotte kernel provides multiprocessing, mechanisms for scheduling, storage allocation, and process migration. The main design decision for the Charlotte operating system is to keep the kernel small, efficient, concise, and easily implemented. Therefore, only those services are included in the kernel (such as IPC and process control) which are essential to the entire system. All other services are implemented as utility processes outside the kernel.

The main disadvantage of Charlotte is the file server design. A weak protection scheme and an unusual directory handling service make it rather difficult to use in practice.

Description

Charlotte is implemented as a new kernel which runs on the Crystal multicomputer, and on a collection of 20 VAX-11/750 connected by an 80 Mbit/second Proteon token ring LAN. It is written in a local extension to Modula except for the startup code which is written in assembler and the interprocess communication code which is written in C.

A software package called the *nugget* resides on every node. Nugget enforces allocation of the network among different applications by virtualizing communications, disks, and terminals.

Charlotte provides a unique form of IPC. A *link* is a capability-protected connection between two processes. Processes address each other by presenting these capabilities (*link-Ids*). Information about the location of a process is hidden inside the kernel. The kernel maintains a link table to locate a particular link. The process-level communication is non-blocking (asynchronous) as well as synchronous and unbuffered. Unbuffered means that a message is not transmitted until the receiver has provided enough space to handle it. The IPC scheme described above is implemented by means of a finite-state automaton for each link. Processes under Charlotte do not share memory.

Any client can gain access to only a subset of the files. If a request is outside of its jurisdiction, a forward to the appropriate remote file server is initiated. The link to the client is part of the forwarding message.

In Charlotte, each file has a unique full path name. Protection is implemented on a file-by-file basis, not by evaluating the rights in directories on a particular path. Therefore, to delete a file, it must be removed from the server on which it resides as well as from the directory. Renaming of directories is not supported.

Failure handling in Charlotte takes an optimistic approach. If a machine fails, it is assumed that within a small amount of time all kernels will discover this and send a *link destroyed* message to all local clients. Replication of resources is not supported. As mentioned above, process migration is part of the design decisions.

Miscellaneous

OTHER ISSUES: In 1984, an implementation of Lynx (Sect. 8.38) for Charlotte was completed. Lynx is a distributed programming language designed at the University of Rochester.

STATUS: Active and in productive use.

CONTACT: Raphael A. Finkel, Dept. of Computer Science, Univ. of Kentucky, Lexington KY 40506, USA, or Michael L. Scott, Dept. of Computer Science, Univ. of Rochester, Rochester NY 14627, USA

e-mail: scott@cs.rochester.edu

REFERENCES: [113] – [117]

4.4 DEMOS/MP

Main Goal

The main goal of the DEMOS/MP project was to provide a basis for various experiments in distributed systems within an easily modified and well structured system. Access and location transparency should be provided and failures of nodes and the network should be handled transparently.

Advantages

The system is simply organized and structured. The two common forms of communication, short messages and bulk data transfer, are well supported. The system tries to improve reliability by a recovery mechanism, but this mechanism and the global name service are centralized. Workstations that are most commonly used nowadays, like SUN or VAX, are not supported.

Description

DEMOS/MP is a distributed version of the DEMOS operating system that has been operational on various architectures including a Cray 1, and is currently running on Z8000 processors connected by a local area network. A simulation was running on a VAX under UNIX.

All communication between processes is done by messages which are sent over one-way message channels called *links*. There are two types of messages, guaranteed and unguaranteed. Guaranteed messages have an at-most-once semantics achieved by an end-to-end acknowledged window protocol. The transfer of large amounts of data is facilitated by a special protocol that uses so-called *link data areas* which can be accessed directly.

All links to a process are maintained by its kernel. A link can be considered as a global address that contains the identifier of the creating machine, a local unique Id and the *last known machine-Id*. Since processes can migrate, the last field is used to localize the current host of a process resulting sometimes in a worse performance. Links are published by a name service, called *switchboard*. Each process can register at its local switchboard and make requests there which are forwarded, if necessary, to the central global switchboard. This leads to a two-level naming hierarchy.

Links are protected objects and provide much of the same function as capabilities.

A recovery facility called *publishing* is installed which is able to recover a process in a completely transparent manner. A centralized *recorder* stores all messages that are transmitted, as well as checkpoint and recovery information. After detection of a failure, the recorder may restart all affected processes from checkpoint. All messages that were sent after the time of the checkpoint are resent to the process while ignoring duplicate messages.

Miscellaneous

OTHER ISSUES: DEMOS/MP provides virtual memory with paging across the network. In this way diskless workstations can be used. Special links provide control of processes across machine boundaries.

STATUS: Dormant – not in use, not under development.

CONTACT: Barton P. Miller, Computer Science Dept., Univ. of Wisconsin-Madison, 1210 West Dayton Street, Madison WI 53706, USA

e-mail: miller@cs.wisc.edu

REFERENCES: [118] – [121]

4.5 DIOS

Main Goal

DIOS is a distributed operating system developed at INESC, Lisboa, Portugal. It was part of the SMD project. DIOS offers a totally distributed UNIX-like file system. The file system is location and access transparent, and independent from the native operating system.

DIOS does not replace the existing operating systems, but instead extends their functionalities. Thereby, DIOS is a generic and highly portable implementation.

Advantages

DIOS was developed to integrate heterogeneous operating systems and required the exploration of hardware mechanisms to support interprocessor message passing in a network environment. Thereby, it uses the same distributed approach as for communication and synchronization inside a closely coupled multiprocessor.

The main disadvantage of DIOS is its elaborate programming style.

Description

In a classic client/server model, DIOS was implemented on bare machines and on top of the commercial UNIX System V. Currently, DIOS runs on mono- and multiprocessor workstations connected through a 10 Mbit/second Ethernet. DIOS was initially conceived with the SMD multiprocessor as a target architecture. However, it runs on VAX-750, on AOS/VS running on a MV8000, and on several UNIX V machines, too. The architecture of DIOS is based on a small kernel which underlies the communication facility.

All system services have globally known local identifiers, which are identical at all sites. The send-and-receive-reply service allows the implementation of synchronous communication. This approach is similar to the RPC technique. Asynchronous communication is also provided. Processes in DIOS do not share address spaces, and communicate exclusively by message exchange.

The file server implements a complete and UNIX-like file system, with all UNIX semantics.

The name manager *(NM)* uses a distributed database that controls the mapping of names to identifiers. NM is implemented through the interaction of name server processes, one at each site. If a name cannot be locally resolved, all remote name servers are interrogated with a broadcast datagram. A monitor maintains the consistency of the database entries.

Security, availability, and failure handling are not addressed.

Miscellaneous

OTHER ISSUES: In parallel with the work on DIOS, a multiprocessor workstation has been developed.

STATUS: DIOS is in everyday use at INESC and other Portuguese companies.

CONTACT: J. Alves Marquez, or J.P. Cunha, or A. Cunha, INESC, Rua Alves Redol N9, 1000 Lisboa, Portugal

REFERENCES: [122], [123]

4.6 Distributed Academic Computing Networking Operating System (DACNOS)

Main Goal

The Distributed Academic Computing Networking Operating System (DACNOS) is part of HECTOR, a joint project of IBM Germany and the University of Karlsruhe, Germany. The main design objectives are preservation of investment (which means that DACNOS must appear as a compatible add-on to existing systems), node autonomy, convenience to the user, i.e., transparent access to remote resources as well as openness and portability.

Advantages

DACNOS is one of the few attempts to provide transparency on top of heterogeneous operating systems. The DACNOS prototype has been operational for only a short time, so that not much can be said about performance or feasibility of the system in an academic environment.

Description

The prototype has been implemented under three types of operating systems: PC DOS on IBM PC AT, VM/CMS on IBM S/370 and VMS on DEC VAX.

A *Global Transport* component (GT) interconnects multiple vendor network architectures, for example SNA, and provides a reliable transport service, which conforms to the OSI layer 4 service interface. The GT transfers datagrams that are reliable by establishing a connection to the target transparent to the user. The *Remote Service Call* (RSC) operations provide an RPC-like mechanism. The operations are available as calls in a programming language. In addition, if necessary, they perform conversion of data.

Global unique names are achieved by concatenating hierarchical domain names and local names within these domains. The *Distributed Directory System* administers these names and locates remote objects by asking a subset of servers which, if necessary, subsequently forward the query.

The *Remote File Access* (RFA) component of DACNOS defines a unified, homogeneous file system on top of the heterogeneous local file systems. The global file names are sequences of identifiers separated by periods. Files become visible in the network by publishing. These files (or file sets) must be mounted and have to be attached to a virtual local file name to guarantee transparent access.

Each user has to be authenticated by the authentication server of his home domain. A server keeps access control lists for all its objects. If a user requests an operation on an object, the server checks the authorization with the help of the server in the user's domain. For this purpose, one authentication server must exist in each domain, installed under a fixed logical node name.

Concurrent access to files is synchronized in such a way that a writer gets a new version which replaces the original file as soon as the writer closes the file. In the meantime multiple readers may read the old version.

Miscellaneous

OTHER ISSUES: The *Remote Execution Service* selects execution sites for running programs remote, can control the running program, and provides access to resources that belong to the environment of the originating unit.

STATUS: The project has been completed since April 1988, and a prototype has been installed. An evaluation of DACNOS is planned in a realistic environment.

CONTACT: H. Schmutz, IBM European Networking Center, Postfach 103068, D-6900 Heidelberg, Germany, or O. Drobnik, Univ. of Karlsruhe, Postfach 6980, D-7500 Karlsruhe, Germany

REFERENCES: [124] – [139]

4.7 DUNE

Main Goal

DUNE is a distributed operating system that provides a single UNIX-like environment, enhanced with process migration and a distributed file system, over a collection of cooperating processors. Its design demonstrates that efficient communication can occur over disparate media without resorting to common protocols that can preclude using unique features of the hardware.

Advantages

DUNE successfully hides the multiprocessor nature of its design from casual users, who are unaware of the processor on which their programs are running or where their files physically reside. Automatic load balancing facilities continuously relocate processes to equalize overall system load. The singly-rooted hierarchical file system space presents a common view to the user, regardless of user identity or process location.

The main disadvantage of DUNE is its reliance on a root processor, which represents a single point of failure.

Description

DUNE is implemented as a new kernel design, rather than a layer on top of a UNIX kernel. As such, it has direct access to important system control blocks for efficiency and flexibility. DUNE runs only on Motorola 68000 family single-board computers.

The interprocess communication in DUNE is procedure-based; the underlying messages, which remote servers use to queue requests, can be transmitted over many media but are unecessary for all requests serviced on the same processor. A hint mechanism, built into the procedure call, informs low-level communications software of optimizations specific to a given medium: Processors that share a backplane interface communicate without copying data, by manipulating their memory management configuration. Processors connected by a token ring interface exchange data using a protocol that takes advantage of the immediate successful-transmission indicator available from the hardware. All communications interfaces look identical to higher-level system software, permitting automatic rebinding of data paths when system objects migrate to other locations.

External objects are named via the hierarchical file system; host/object identifiers identify internal system objects. Process identifiers remain constant, even when a process migrates to another processor.

DUNE does not explicitly address security, assuming that all kernels are trusted and reliable. Globally unique user identifiers control file access.

Miscellaneous

STATUS: DUNE development is essentially complete. There are no current plans to enhance the system to more modern hardware.
CONTACT: Marc Pucci, Bell Communications Research, 435 South Street, Morristown NJ 07960, USA
e-mail: marc@bellcore.com
REFERENCES: [140] – [142]

4.8   DUNIX

Main Goal

DUNIX integrates several computers connected by a local area network into a single UNIX machine providing access and location transparency. In addition, the complexity of the kernel is supposedly reduced.

Advantages

DUNIX meets its goal to provide a distributed UNIX with acceptable performance and increased availability, but almost nothing is done to improve reliability and consistency.

Description

The system is implemented on VAX-machines connected by an Ethernet and consists of a complete new kernel. The underlying hardware has to be homogeneous. Almost all 4.x BSD binaries run without recompilation.

The communication system uses its own protocol which provides at-most-once semantics. For communication with other systems, it implements the standard TCP/IP protocols.

DUNIX has a single global name space for processes and files as in UNIX. Additionally the file system is partioned into groups with a *superroot*, which is named '...', so each group has its own root directory ('/') providing the usual naming mechanism. However, the system can contain several copies of its vital files, for example /bin. There is not necessarily any correspondence between physical computers and sub-trees. Each '/dev' directory names all of the system devices. Each symbolic name is mapped to a *universal file name* that has a fixed size and contains the host id, an index into the primary memory file table and a unique identifier.

The protection scheme is the same as in UNIX.

The availability is improved, but no measures are taken to improve reliability.

Miscellaneous

OTHER ISSUES: Disks can be switched manually from a crashed computer to a running one with all files afterwards accessed in the same way.

STATUS: An installation is running at Bell Communications Research; some of its ideas were explored in the MOS system (Sect. 4.14).

CONTACT: Ami Litman, Bell Communications Research, 435 South Street, Morristown NJ 07960, USA

e-mail: ami@bellcore.com

REFERENCES: [143], [144]

4.9   Freedomnet

Main Goal

Freedomnet was developed to extend UNIX semantics beyond machine boundaries, allowing UNIX-based systems to become a distributed computing system capable of transparently sharing files, peripherals, and CPU cycles across a network.

Advantages

Freedomnet software is added to standard UNIX kernels and extends their particular semantics with access and location transparency. For example, "dial-up" programs such as *tip* or *kermit* can use modems that are physically connected to remote computers. Programs may be executed on any system able and willing to run them. For example, a user may specify their login shell to be one that is for a different processor type and/or one that is resident on a remote system. If necessary the program will be copied/executed on a processor of the appropriate type. Command line editing, command execution, and job control will work properly despite any operating system differences. During execution the process may migrate to another computer.

The main disadvantage of Freedomnet is that it is currently unsupported by any computer vendor, so kernel problems must be debugged with help from Freedomnet technical support.

Description

Freedomnet runs on most major versions of UNIX including DEC VAX and RISC (Ultrix, 4.3 BSD), Hewlett-Packard (HP-UX), IBM (AIX), SUN 680X0 and SPARC (SunOS), 386/ix, SCO UNIX, Sequent (Dynix), Pyramid (OSx), UNISYS (System V.2, V.3), Convex, Encore (UTX/32), Amdahl (UTS), and NCR (System V.2). Freedomnet is a system call distribution layer that resides within the kernel at the system call interface. It is added to the standard kernel in much the same way as a device driver.

A remote system call (RSC) protocol is used for communications between the Freedomnet layers on the different systems. The role of the Freedomnet layer is to intercept system calls and interpret them to see if remote objects are being referenced. If not, the system call is passed on unchanged to the local kernel. If so, an appropriate remote system call is constructed and passed to the remote machine where the system call is executed by the remote kernel. The result is passed back to the local machine and relayed back to the invoking program.

Each client process has access to remote objects (files, devices, etc.) through a stateful remote server process that provides at-most-once semantics. When called upon by a client process, a server manager running continuously on the remote system and listening at a well-known port creates a unique remote server for the process. The client process accesses the full range of UNIX operations through this server process. Even though a server process resides in a remote machine, it is, in effect, an extension of the local user process.

Freedomnet extends the UNIX directory tree into a *supertree* (a generalization of *superroot*, Sect. 4.15) of interconnected systems. This is the preferred naming model. Additionally, or alternately, any remote directory or file can be mounted onto the local file system, providing "fine-grained" remote access.

Mapping of remote user-Id or group-Id to local user identifications is provided. Client authentication and communication security depends on a secret-key mechanism that will utilize Kerberos 5.0 (Sect. 6.1) as it becomes available.

A server crash causes loss of all server state. Freedomnet clients track some of the server state (e.g., working directory) and can usually reestablish the server. Some server state (open devices) seems impossible to track. A client crash leaves orphaned server processes which are periodically detected and terminated. Freedomnet does not itself provide stable storage, but distributes system calls that do (e.g., IBM AIX *commit* and *abort*).

Freedomnet implements process migration as a generalization of remote execution. The process execution is suspended during the migration. Migration must be explicitly requested, dynamic process migration remains unimplemented.

Miscellaneous

Freedomnet performance is comparable to that of NFS (Sect. 2.8) for ordinary file system activities. Communication traffic can in many cases be reduced with remote execution. A *pickfast* routine reports the currently fastest suitable processors for a given task, and is currently being used to determine execution sites for spreadsheet calculators, relational database systems, parallel make, and of course parallel Mandelbrot set generation.

STATUS: Freedomnet is an established commercial product available from the Research Triangle Institute.

CONTACT: Thomas Warren, Research Triangle Institute, PO Box 12194, RTP, NC 27709, USA

e-mail: wtw@rti.rti.org

REFERENCES: [145] – [147]

4.10 HERMIX

Main Goal

HERMIX is a distributed operating system, in which the system appears to the users as a traditional uniprocessor system. The system software is distributed among all nodes. It is highly transparent in terms of location and access. The structure of the system is modular. Only the basic services such as local communication, low-level process management, low-level memory management, and interrupt handling are provided by the kernel. The rest is being offered by servers. This has been chosen to make it possible for higher services such as high-level naming, replication, and failure handling to be transparent.

Advantages

The design, although extremely well structured, will not lead to high performance. From the literature today, it is clear that short cuts for the often occurring "cases" are necessary in order to build systems with higher performance.

Description

HERMIX is written for a Motorola 68000, but with heterogeneity in mind. It is a completely new kernel, built from scratch.

The HERMIX interprocess communication scheme is general and uniform. It combines the well-known RPC, which optimizes on low latency, with a byte-stream protocol, which optimizes on throughput, thus providing one general communication mechanism. Furthermore, communication is type-safe, and supports heterogeneous environments. The communication system also offers a simple but flexible fault model, which forms the basis for building fault-tolerant systems.

The communication system provides security, authentication, recovery of crashed entities, integrity, and privacy. Moreover, it provides facilities for transactions.

Process migration is not applicable.

Miscellaneous

STATUS: Research on the HERMIX system started around 1983; two PhD's and a number of reports and articles handled the topics. A limited prototype was built on top of bare hardware. As the designers had not enough man power to add the necessary device drivers and tools to their prototype in order to make it a real working system, they decided in 1987 to work on top of UNIX for further research.

The two PhD's related to HERMIX handled structural aspects and communication. Current research includes system support for object management in heterogeneous distributed environments, and load balancing in distributed systems. This research has of course been influenced by the experience gained in the HERMIX system, but is not directly related to HERMIX.

In the XENOOPS project the designers now investigate execution environments for object-oriented parallel software on parallel computing systems. Such environments should provide dynamic load distribution.

CONTACT: Yolande Berbers, Katholieke Universiteit Leuven, Dept. Computerwetenschappen, Celestijnenlaan 200 A, B-3030 Heverlee, Belgium

e-mail: yolande@cs.kuleuven.ac.be

REFERENCES: [148] – [157]

4.11   JASMIN

Main Goal

JASMIN was an experimental distributed operating system kernel that provided message passing over dynamic communication capabilities (location transparency). It was in use as a distributed processing research tool at Bell Communication Research.

Advantages

JASMIN was a merely academic approach meant to demonstrate and evaluate some interesting fields in distributed environments.

Description

JASMIN functioned as a client/server model. It consisted of a new operating system kernel with an execution environment of servers and a connection to an UNIX host computer. The JASMIN kernel provided IPC, tasking and scheduling, whereas other services were provided by server tasks separate from the kernel.

In JASMIN, communication between so-called *tasks* was via messages sent on *paths*. Messages were sent asynchronously. The sender did not block as in the typical client/server model. The kernel guaranteed that messages were reliably delivered in the order they were sent in.

Paths were one-way communication links that granted the holder a capability to send messages, similar to the links of DEMOS/MP (Sect. 4.4) or Roscoe (Sect. 8.56).

JASMIN maintained a two-tiered hierarchy of name servers. First, a single global name server (GNS) and second, multiple local name servers (LNS). LNSs had a path to the GNS. LNSs determined whether a server was local or remote.

Miscellaneous

OTHER ISSUES: A distributed database management system was a major application of JASMIN.

STATUS: Not active.

CONTACT: Dan Fishman, Hewlett-Packard Research Lab., Palo Alto CA, USA
e-mail: fishman@hp.com

REFERENCES: [158] – [161]

4.12 LOCUS

Main Goal

LOCUS was developed to provide a distributed and highly reliable version of an operating system to the huge UNIX community. LOCUS tools support an automatic conversion from stand-alone UNIX nodes to network-wide connected LOCUS nodes simultaneously offering a high degree of transparency, namely location, concurrency, (optimistic) replication and failure transparency.

Advantages

LOCUS' main advantage is its reliability and availability feature supporting automatic replication of stored data, with the degree of replication dynamically under user control. Due to a majority consensus approach within partitioned subnets to build up a temporary still-connected partition, the system remains operational with a high probability. This feature implies that sometimes out-of-date data will be modified resulting in an inconsistent global state. LOCUS helps the users to examine inconsistent data.

In the commercial version of LOCUS a simple primary-site replication scheme is used.

The tradeoff between performance, remaining operational as well as consistency leads to LOCUS' disadvantage: Even when a merge-protocol frees a user in most cases from consistency checks after network partitionings, sometimes user interaction can still be required in order to get back into an up-to-date global consistent state. Through the use of a centralized synchronization site and fully-replicated logical mount tables, it does not scale well.

Description

LOCUS is a step-by-step modification and extension of the UNIX operating system compatible with the BSD and System V branches. Load modules of UNIX systems are executable without recompiling. Tools exist for an automatic conversion to LOCUS by exchanging the UNIX kernel with a so-called LOCUS kernel and by reformatting an existing UNIX file system without loss of information. The 1983 version of LOCUS supports DEC VAX/750, DEC PDP-11/45 and IBM PCs connected by Ethernet or Token-Ring LANs. The communication between different LOCUS kernels uses a point-to-point connection (VC, virtual circuit) via specialized remote operations protocols.

In spite of replication and caching, full UNIX semantics are emulated. Alternatively, file locking is provided.

A single tree structure spanning all nodes of the network includes virtually all the objects of the file system. All objects possess a global unique name in a single uniform and hierarchical name space, identified by a path name (character string). There is only one logical root in the entire network. The path names do not include information about the location of objects. So-called *file groups* are self-contained subtrees within the naming hierarchy.

Security in terms of intruder defense or secure data transmission plays a minor role and the mechanisms are comparable to the ones implemented in UNIX.

File replication is made possible in LOCUS by multiple physical containers for each logical file group, and gaining access to a replica is done according to a certain protocol: A centralized synchronization site, the so-called CSS (*current synchronization site*), decides whether a request from a US (*using site*) is possible or not (is a copy available?), and determines the actual SS (*storage site*) needed to satisfy this request. An atomic commit using shadow copies of updated pages provides consistency in cases of crashes or network failures. Nested transactions can be made. Token passing for distributed operations on the file system permits serialization.

Miscellaneous

OTHER ISSUES: Remote tasking is supported by two new (remote process) system calls, *migrate* and *run*.
STATUS: Originally a research effort, at UCLA in the late 1970s, LOCUS nowadays has been transferred to a commercial venture: LOCUS Computing Corporation, Santa Monica CA, USA
CONTACT: Gerald Popek, Dept. of Computer Science, Univ. of California, Los Angeles 90024-1596 CA, USA
e-mail: popek@cs.ucla.edu
REFERENCES: [162] – [172]

4.13   MACH

Main Goal

MACH was designed with the intention of integrating both distributed and multiprocessor functionality. In addition full binary compatibility with UNIX 4.3 BSD should be given by preserving a simple, extensible kernel that runs most of the services, for example file service, in user-state code. It is an extension of the Accent system (Sect. 4.1) to multiprocessors and other types of hardware, providing the same classes of transparency and adding some further features.

Advantages

MACH is a good environment for distributed and parallel applications because of its variety of inter-process communication possibilities, ranging from shared memory to efficient message passing for uni- and multiprocessors. Moreover it provides a good chance for the huge UNIX community to change to a real distributed system without giving up their convenient services. Although MACH is more extensive, its performance is in line with the Berkeley UNIX. On the other hand not much work is done on built-in failure handling and reliability improvements.

Description

MACH's current implementation runs on all VAX uni- and multiprocessors, SUN 3s, IBM RT-PCs, the Encore Multimax with up to twenty NS 32032 series processors, the I386 and the Mac II. As mentioned above, it is fully binary compatible with UNIX 4.3 BSD.

MACH uses a simplified version of the Accent IPC facility because lightweighted processes (threads) exist that are able to handle some forms of asynchrony and failures more efficiently. Unlike BSD, there is one interface, the port, which is consistent with all resources. See Accent (Sect. 4.1) for more information concerning communication, naming and protection.

In addition, MIG, an interface specification language, enables interfaces between clients and servers to be specified and generates remote procedure call stubs. These interfaces are able to perform run-time type-checking and deliver enough information to the network servers to make data-type conversions that exchange between different types of hardware.

Many of the missing reliability issues are included in a system called Camelot (Sect. 8.11), which relies on MACH. Camelot is available from Mt. Xinu of Berkeley, CA.

Miscellaneous

OTHER ISSUES: It is possible to share memory between multiple threads (the entities of control) and tasks (the entities of resource allocation) in a highly machine-independent manner, providing good performance by the use of copy-on-write virtual memory. An ADB-like kernel debugger is available.

STATUS: MACH is running at Carnegie Mellon University, supporting the Andrew distributed file system (Sect. 2.2). Furthermore, it is the basis for a variety of other approaches, such as Agora, Avalon, Guide, MARUTI, and Synthesis (Sects. 8.2, 6.2, 5.11, 5.13, and 6.21, resp.). It is free for distribution and is available for export. Development and improvement are still in process and new versions will be available to all users. Universities and research centers are asked to contact Carnegie Mellon University. Corporate sites are asked to contact Mt. Xinu.

CONTACT: Richard F. Rashid, Computer Science Department, Carnegie Mellon University, Pittsburgh PA 15213, USA, or Noele Krenkel, Mt. Xinu, 2560 Ninth Street, Berkeley CA 94710, USA

e-mail: mtxinu-mach.@mtxinu.com resp. mach@cs.cmu.edu

REFERENCES: [173] – [187]

4.14   Multicomputer Operating System (MOS)

Main Goal

The Multicomputer Operating System (MOS) is a distributed UNIX system. It is a general-purpose time-sharing operating system which makes a cluster of loosely connected independent homogeneous computers behave as a single-machine UNIX system. The main goals of MOS include location and access transparency, decentralized control, site autonomy and dynamic process migration.

Advantages

MOS is one of the few systems that attempts to integrate load-balancing philosophies into a distributed operating system. Process migration and a suboptimal load-balancing algorithm are implemented.

Description

MOS has been developed at the Hebrew University in Jerusalem and consists of seven PCS/CADMUS 9000 machines (with M 68010 CPU) interconnected through a 10 Mbit/second Pronet local area ring network. The machines must be homogeneous in order to allow process migration. Diskless nodes are supported.

MOS is an enhanced UNIX V7 kernel supporting all UNIX system calls. Thus, MOS is binary compatible with UNIX. MOS uses the RPC mechanism based on the Unreliable Datagram Protocol for communication.

The file system in MOS is a forest. Each tree is a complete UNIX file system. A single file system is achieved by placing a *superroot* above all trees of the forest. Standard, path named, hierarchical UNIX naming conventions are provided.

Security issues extending UNIX features are not implemented. An outside-kernel software exists that provides replication of storage to increase the system's availability.

No provisions are made for failure handling. MOS does not support atomic file transactions since these features are not supported by UNIX.

Miscellaneous

OTHER ISSUES: The MOS Multicomputer Operating System is a distributed implementation of UNIX providing *suboptimal load balancing* by means of process migration. Each machine stores load information from a fraction of other machines, including itself. Periodically the recent half of that information is transmitted to a machine which is chosen at random. The transmitted information replaces the "stale" half of the load statistics previously known from the receiving machine. Processes always migrate to underloaded nodes and are executed remotely. In addition lightweighted processes exist which can be executed on different machines.

STATUS: A PDP-11 version of MOS has been in operation since 1983, the current version is operational since Spring 1988. Research is going on in performance modeling, reliability, parallel algorithms and distributed applications.

CONTACT: Amnon Barak, Dept. of Computer Science, The Hebrew Univ. of Jerusalem, Jerusalem 91904, Israel

e-mail: barak@hujinix.bitnet

REFERENCES: [188] – [197]

4.15   Newcastle Connection

Main Goal

The Newcastle Connection (or NC) was developed at the University of Newcastle upon Tyne, UK. The NC was the implementation of an architecture for distributed UNIX systems called UNIX United, also developed at Newcastle. This architecture was based on a recursive structuring principle, namely that "a distributed system should be functionally equivalent to the systems of which it is composed". Thus, a UNIX United system constructed with the Newcastle Connection is functionally indistinguishable from a centralized UNIX system at the system call level.

This means that the users of a UNIX United system can access the resources of a remote UNIX system as if they were local; all network accesses are transparent. However, location transparency is not provided.

Advantages

The NC influenced a great deal of the research and thinking about distributed systems at Newcastle and elsewhere during the 1980s. The idea of recursive structuring on which the NC was based proved to be particularly fruitful, leading to a clean separation of concepts in the construction of distributed systems.

The NC was one of the first and most complete attempts at building a transparently distributed UNIX system. It was made available to other academic institutions through an educational license and thus gave many students in the early days of distributed systems the opportunity to study a complete system. It was also made available commercially and was used as the basis of several products.

Indeed, the NC was distributed as part of the standard UNIX implementation on the ICL Perq workstation which was widely used throughout the UK academic community in the early 1980s as the result of the UK Science and Engineering Research Council's Common Base Policy.

The transparent distribution layer provided by the NC is at a higher level of abstraction than the remote filesystem abstraction used by systems such as NFS (Sect. 2.8) and RFS (Sect. 2.9). To the extent possible within the constraints of the UNIX system call interface, the NC provides a complete emulation of a distributed UNIX system at the UNIX system call level. Thus, remote programs are executed remotely rather than locally, and signals and pipes between remote processes are fully supported.

UNIX's flat name spaces for pids, uids, etc., do not generalize easily to a recursively transparent distributed system. Maintaining strict compatibility with the semantics of the various UNIX variants supported whilst supporting heterogeneous interworking between them required compromises to be made and detracted from the research contributions of the NC.

Description

The original implementation of the NC was for UNIX V 7 running on a set of DEC PDP-11 computers, connected together via the Cambridge Ring local area network.

Subsequent implementations added support for other versions of UNIX (including BSD 4.2 and System V), network technologies (Ethernet, Serial line), protocols (LLC1 datagram, UDP/IP), and hardware architecture (VAX, 68000). The NC was designed to be easily portable to a new implementation of UNIX and to support interworking between different versions of UNIX. A protocol-independent network interface was devised to allow the NC to act as a gateway between machines which were otherwise incompatible at the network protocol level.

The NC was implemented as a transparent layer of software emulating the UNIX system call interface which intercepted all the system calls made by a UNIX application and forwarded them to a local or remote UNIX kernel as appropriate. The NC was originally implemented outside the UNIX kernel as an alternative version of the UNIX system library that applications could be linked with to gain access to remote resources. However, implementations of the NC inside the UNIX kernel were also built.

Because it was functionally equivalent to a UNIX system call interface, the NC could be used by a UNIX application or inserted into a UNIX kernel without any modifications to existing source code.

The NC was designed according to the classical client/server model. The role of the client side was to filter out system calls that needed to be redirected to a remote system whilst passing local system calls on to the underlying kernel. The role of the server side was to handle remote system calls directed at it by the NC layer on another UNIX system. The two sides communicated using a reliable RPC built from an underlying unreliable connectionless network protocol. The RPC protocol supported exactly-once semantics with "wait for ever" timeouts to handle system calls that could execute indefinitely, e.g., terminal reads, process waits. This experience of using RPC protocols to implement real distributed systems, warts and all, influenced the theoretical work on RPC protocols being carried out at Newcastle around the same time.

The NC joined UNIX systems together using what was effectively a remote mount. However, the entire name space of a remote system was mounted (subject to the usual access controls) rather than just a subtree. In other words, whole systems rather than disk volumes were mounted. Mounts were intended to be transitive but the correct handling of UNIX semantics in these circumstances requires an additional layer of protocol which was designed but never completely implemented. (A simple recursive invocation of the NC by the server is not sufficient although it is easily implemented.) In practice, this did not prove to be a problem because the naming tree was intended to be symmetrical, making it possible to access any system directly from any other. However, the architecture was more general than this and could scale to support an infinite naming tree, independent of the underlying network topology and system connectivity. In fact the whole question of transitivity and the joining together of distributed name spaces according to the recursive structuring principle was studied at length, both theoretically and in practice.

To the user of the NC, a remote system appeared to be a directory. Since the NC was usually used to connect together autonomous UNIX systems, each system was regarded as equivalent in the naming hierarchy of the distributed system. Thus, it was conventional for the directories denoting remote systems to appear in a parent directory of the local root directory. This parent directory was denoted "/.." and was sometimes referred to as the *superroot*. However, the fact that the superroot existed at all or was replicated across all the machines in the UU distributed system was not fundamental to the NC design but simply the most logical way to name otherwise equivalent UNIX systems. Thus, there was nothing special about the superroot – it was just an extension of the normal UNIX naming mechanisms. Remote systems could just as easily have been accessed from any other part of the UNIX naming tree. On the other hand, full location transparency is not provided since the users are aware of leaving their local file system.

The NC allowed both the current and root directory of a process to be remote. If a pathname passed through or started from a remote system, it was interpreted relative to that system. UNIX identifiers not derived from pathnames (e.g., process Ids, user Ids) were interpreted on the system where the root directory was located (which was not necessarily the same as the system where the process was executing). A special form of remote execution that moved the root directory to the remote machine was provided as an alternative to normal remote execution which left the root directory unchanged. This was useful for remotely executing UNIX utilities such as "ps" and "who" that depended on root-relative Ids.

Access to remote systems in an NC cluster was controlled by the local system administrators. It was possible to specify which remote machines could access your local machine and what permissions users of those remote machines would have on your local machine. A simple user id mapping scheme, transparent to users of the NC, meant that there was no need for a unified password file. Having logged on to his or her local UNIX system, a user could access the facilities of a remote system, subject to both local and remote access permissions, without any further need for authentication. It was up to the administrator of each system to decide how trustworthy he or she considered other machines on the network to be.

Replication, load balancing, fault tolerance and security were considered to be additional capabilities which could be implemented on top of the NC by virtue of the fact that the NC provided a distribution layer rather than incorporating them as essential parts of the NC. Indeed, research was performed into all these areas and several prototypes were built using the NC as a basis. The actual implementation of the NC itself was fault-tolerant in the sense that the RPC protocols used provided exactly-once semantics in the presence of client and server crashes with orphan killing. However, servers were stateful and thus server crashes were not transparent to clients in the way that they are with NFS.

Miscellaneous

STATUS: The original implementation of the NC was for UNIX V 7 but later releases of the NC tackled the heterogeneity problem and supported interworking between different variants of UNIX, including V 7, System V and BSD 4.2. System calls which were not provided directly were emulated and translations between internal data formats such as directory structures were made. The NC was also used as a basis for other experiments in heterogeneity involving the connection of UNIX and non-UNIX systems.

The recursive structuring philosophy on which the NC was based led to a clean separation between issues concerned with the construction of a distributed system and issues concerned with exploiting that distribution to provide increased reliability, security, availability or whatever. Several complementary mechanisms to provide such capabilities were designed to exploit the facilities provided by the NC in a layered fashion.

Another area which was explored extensively using the NC as a basis was that of naming – in particular, the extent to which local name spaces based on locally unique identifiers could be joined together recursively to form larger name spaces. The UU architecture on which the NC was based allowed for the indefinite expansion of the UNIX name space across an arbitrary network topology.

The NC achieved limited success as a commercial product and forms the basis of Research Triangle's Freedomnet (Sect. 4.9) which is still marketed today. However, NC specific research is no longer active at Newcastle and all development of the NC as a product has ceased.

CONTACT: B. Randell, L.F. Marshall, R.J. Stroud, Computing Laboratory, University of Newcastle upon Tyne, Claremont Road, Newcastle upon Tyne NE1 7RU, UK
e-mail: {brian.randell, lindsay.marshall, r.j.stroud}@newcastle.ac.uk
REFERENCES: [198] – [207]

4.16   PULSE

Main Goal

PULSE was a distributed operating system written partially in Ada. The main aspects of the project were the design of a distributed file system that could support standalone operation of workstations, as well as the evaluation of Ada's suitability for distributed applications.

Advantages

The system provided a good environment for distributed Ada programs. The use of the prototype is circumstantial because two types of machines are required, one for developing and one for running applications.

Description

PULSE was run on several LSI-11/23 that were connected by a Cambridge Ring. All development was done on a VAX, i.e., cross-compiled and downloaded, since no Ada compiler existed for LSI. The kernel was written in C.

The PULSE IPC provided a virtual circuit and guaranteed only that messages received would be in good condition and in sequence, but did not guarantee delivery. Messages could be sent synchronously, asynchronously or RPC-like.

The filesystem had a UNIX-like hierarchical structure. Files were kept in *volumes* that had unique numbers. Location was done by associated global *i-numbers*. Each file server maintained a list of locally on-line volumes. If a file was remote a request was broadcast.

The protection scheme was the same as in UNIX.

Files and directories were replicated and updated with a *primary copy* mechanism. The duplicates were updated the next time they were referenced. For this purpose, version numbers were held. A user could control replication. When a client referenced a file, the file server attempted to find a copy in the following order: local master, local duplicate, remote master, remote duplicate.

Miscellaneous

OTHER ISSUES: Several software tools, such as shell, ed, etc., were implemented in Ada.
STATUS: A prototype was implemented. The project is now finished and the results have been published as a book. The designers are currently working on a new distributed operating system called Wisdom (Sect. 4.23).
CONTACT: A.J. Wellings, Dept. of Computer Science, Univ. of York, York, UK
e-mail: andy@minster.york.ac.uk
REFERENCES: [208] – [213]

4.17   QuickSilver

Main Goal

QuickSilver, developed at the IBM Almaden Research Center, is a client/server structured distributed system which uses atomic transactions as a paradigm for recovery management. This way the system is access, location and failure transparent. Properly written programs should be resilient to external process and machine failures and should be able to recover all resources associated with failed entities. The main recoverable service fully operational is the Distributed File Service.

Advantages

The system provides highly reliable services with little loss of performance, for example the recovery management overhead is low.

On the other hand, nested transactions and language directed facilities are not available for defining and using recoverable objects.

Description

QuickSilver consists of a new operating system kernel, running on IBM RT-PCs interconnected by an IBM token ring. QuickSilver also runs on the 370 architecture under VM/CMS and has been ported to the RS/6000 family.

The inter-process communication can function synchronously, asynchronously (with a returned request-Id for later use) or by way of messages. Most commonly, a request-response protocol is used, in which requests are neither lost nor duplicated, data is transferred reliably, and in case of a server crash, errors are reported to any clients involved (at-most-once semantics). Each service has a global unique address which is private or otherwise made public by registering with the name service.

Each resolution of a file name in the QuickSilver distributed file system is performed in the context of a specific user's file name space. The universe of users is partitioned into domains, each having a user location and identification service. User names are unique within a domain, so that the (domain, user name)-pair leads to a global unique name space. Each file server has a recoverable copy of all user indices and quotas. The Distributed File Service has transactional behavior for all its metadata. Moreover it supports multi-node atomic actions by collaborating with the Transaction Manager in two-phase commit.

The only authentication is made at login. Afterwards, the kernel and the communication system are trusted as in a local system.

To cope with failures, each IPC, on behalf of a transaction, is tagged with a global unique transaction identifier. Transaction Managers (TM) are responsible for commit coordination by communicating with servers at their own node and TMs at other nodes. The commit coordinator can migrate to the subordinate or, in the case of more than one subordinate, be replicated to all subordinates. A Log Manager serves as a common recovery log both for the TM's commit and the server's recovery data. Servers are declared stateless, volatile or recoverable. No serialization is provided. This remains the responsibility of the servers managing the serializable resource.

Miscellaneous

OTHER ISSUES and STATUS: A network transfer protocol, OBJ, has been designed and implemented. It seeks to minimize the number of messages needed for a request by providing higher-level semantics to the end points of a communication.

A Deadlock Detector is going to be developed that can detect global deadlocks and resolve them by aborting offending transactions. The system is running in daily production use at IBM Almaden. Experiments are being made with other applications, including a messaging facility and a mail store-and-forward system.

CONTACT: Luis-Felipe Cabrera or Roger Haskin, IBM Almaden Research Center, 650 Harry Road, San Jose CA 95120-6099, USA

e-mail: cabrera@ibm.com or haskin@ibm.com

REFERENCES: [214] – [216]

4.18   Research Oriented Distributed Operating System (RHODOS)

Main Goal

RHODOS stands for Research Oriented Distributed Operating System. It is developed as a testbed to study design issues of distributed operating systems (generalized interprocess communication, load balancing, object naming, protection, process and file migration, authentication and communication security, heterogeneity, and object replication) and their influence on the overall system performance. It provides network transparency, in particular access, location, (migration), and in the future replication and failure transparency.

Advantages

RHODOS is a new distributed operating system, based on the two years of extensive research and design, and detailed analysis of existing experimental distributed operating systems. The system is built as a highly portable and integrated distributed operating system for heterogeneous workstations with a very modular process-based (as against monolithic) kernel with a flexible communication subsystem having a generalized set of communication semantics. Process migration is used to support load balancing, and the communication subsystem has special built-in functions supporting this. RHODOS can use different naming and resource protection schemata (capabilities or User Access Lists). Message encryption and user authentication is provided and system names have special support for user authentication.

Description

RHODOS is written in C++, and is currently midway through its implementation on a LAN of four SUN 3/50 workstations. It is based on a communication subsystem that provides message passing and connectionless remote procedure calls with at-most-once, at-least-once, and exactly-once qualities of service, with user-transparent server liveness checking. This in turn is achieved with a reliable connectionless datagram transport protocol (RHODOS Reliable Datagram Protocol) which interfaces with IP, and thus allows LAN and WAN communication.

Interprocess communication is done through *ports*. Each process has a unique port, a "hidden port" used for RPCs, and possibly other ports. Servers advertise their service ports to a name server to provide well-known services.

The kernel is broken into several kernel server processes, with a small non-process nucleus which handles hardware interrupts, local message passing and context switching. The kernel servers form a team of *mediumweight* processes, sharing a common data space, whilst having individual text and stack spaces. These use interprocess communication to perform the services of the RHODOS kernel. The policies which drive the system are instituted in a set of manager processes, which are the clients of the kernel servers. This modularity allows the fast and flexible changing of the RHODOS environment.

Memory management is based on an abstract object, the *space*, which has several machine-independent attributes, allowing inter-machine portability. Each process has at least three spaces (text, stack, data); data and stack can be varied in size, and data may be shared amongst several processes. Files may be mapped in as spaces, allowing direct memory access to the file. The memory manager provides LRU paging, and lazy page allocation for efficient memory use. File mapping is used to migrate a process' spaces via a common file server.

Three levels of naming are provided: physical names (known only by the kernel), system names (known by user processes and the kernel), and user names, attributed names which are used by the user. A name server is provided to map from user names to system names, and also provides white pages services, yellow pages services, and hierarchical names for efficiency. RHODOS provides both location and access transparency at the user name and system name level. Different naming resolution and location strategies will be implemented and tested.

Resource protection is provided through either capabilities, User Access Lists, or a mixture of them. Communication security is provided using encryption and authentication, and the naming facility has built-in support for user authentication.

Process migration is performed by the migration manager in conjunction with the interprocess-communication manager and is driven by the load balancing server. The "which, where and when" decisions made by the load balancing server are made based on data (computational load and communication patterns) gathered by another kernel manager, the collector. Negotiation and actual migration policy can be easily changed to support such things as: negotiation, load data exchange, centralized or distributed data collection, etc.

Miscellaneous

CONTACT: Andrzej Goscinski, Dept. of Computer Science, University College, Univ. of New South Wales, Canberra, ACT 2600, Australia
e-mail: ang@csadfa.cs.adfa.oz.au(@uunet.uu.net)
REFERENCES: [217] – [224]

4.19   Saguaro

Main Goal

Saguaro is a distributed operating system which attempts to strike a balance between hiding the underlying hardware (for example, providing location transparency) and allowing users to take advantage of it (for example, placing replicas of files on different disks).

Advantages

Saguaro is a synthesis of new mechanisms and variations on ones from existing systems and provides various ways in which users can exploit the multiple machines offered to enhance performance and availability.

The entire system was never finished, so it remains to be shown whether these concepts are feasible and perform well.

Description

Saguaro was an attempt to write an operating system in the high-level distributed programming language *SR*. A prototype implementation of the file system portion was done on a network of eight SUN workstations. It was implemented on top of the UNIX kernel using the IBIS system (Sect. 2.7).

Processes communicate by synchronous calls or asynchronous send. Input and output of different commands can be connected by *channels* the functionality of which is a superset of UNIX pipes. Each channel has one write port and one or more read ports.

Files in Saguaro are organized into a single, hierarchical *Logical File System* (LFS) that is mapped onto one or more *Physical File Systems* (PFS). Any file or directory can be placed on any PFS, either explicitly by adding the PFS name or by default in the same PFS as its parent directory. Files are located by scanning the path name. If a file along the path is inaccessible, a broadcast is made to the manager of every PFS which has access to the logical files stored on its PFS by a *virtual root*. In this way every file is accessible as long as the PFS on which it is stored is accessible.

Security was not a research topic.

Files can be replicated by the use of special calls which have the file names of the replicas as parameters. The first call allows direct access to each replica and the second provides transparent access. The system tries to keep replicas consistent, returning error messages if this is not possible.

Miscellaneous

OTHER ISSUES: In Saguaro, all data is typed and the system performs type checking on command invocation and file access. The basis for this is a data description language (UTS). Servers are created dynamically and placed on an adequate host by a load manager. Physical resources are controlled by managers that are created during system initialization.

STATUS: Active research on Saguaro stopped in 1988.

CONTACT: Richard D. Schlichting, Department of Computer Science, University of Arizona, Tucson AZ 85721, USA

e-mail: rick@cs.arizona.edu

REFERENCES: [225] – [231]

4.20   Sprite

Main Goal

The Sprite operating system is designed for a set of possibly diskless workstations, connected by a local area network. The system is compatible with 4.3 BSD UNIX and is modeled as a set of services including a shared hierarchical file system and remote program execution with paging across the network. Full UNIX file semantics are provided, including locking and consistent access. Furthermore, it provides access, location, and concurrency transparency. It provides semi-automatic transparent process migration to permit parallel execution of applications on idle hosts.

Advantages

The Sprite file system has better performance than comparable file systems, such as NFS (Sect. 2.8), due to better client caching. In addition it copes better with concurrent writes. Process migration commonly provides speedups in the range of 3-6 for parallel compilations and linear speedups for CPU-bound simulations.

The main disadvantage at present is that it is hardly portable because the communication protocol is only implemented for one type of local area network.

Description

Sprite is an operating system kernel running on hosts connected by an Ethernet. Initially, Sprite ran on SPUR, a multiprocessor workstation developed at Berkeley. Today, the system runs on DECstation 3100, SPARCstation 1, SUN 3, and SUN 4 workstations, as well as the Sequent Symmetry. The interface for user processes is much like that provided by UNIX. Remote links to files on hosts running UNIX are possible as well.

Its communication system is based mainly on an asymmetric kernel-to-kernel Remote
Procedure Call, which in turn is currently based on a special purpose protocol for an
Ethernet. The internet protocol is supported, but only used when communication is
required with hosts that are not on the local net. In order to reduce state information
and allow implicit acknowledgements by getting replies or sending new requests, a set of
clients is associated with one server process by the use of so-called *client channels*. This
way at-most-once semantics can be achieved by saving copies of the last reply message
and using sequence numbers as well as explicit acknowledgements and retransmissions
in case of timeouts. A large message can be sent in one RPC by dividing it into fragments that allow partial retransmissions of lost fragments. Built-in delays that prevent
overflowing by slow servers are implemented. If a channel is idle, the server processes are
freed in LRU fashion. If there is no server available at the beginning of a transmission
over a channel, the message is discarded. It is possible to broadcast an RPC, receiving only the first reply. However, broadcast RPCs are limited to prefix locations and a
special gettimeofday, used at startup.

Sprite provides a hierarchical file system in which subtrees can be managed by different servers. Instead of a name service, each client has a prefix table which contains
domain names, e.g., directories, the address of the server who manages this domain, and
a short identifier of this directory. This table is constructed by using a broadcast and
updated automatically as the configuration of servers and files changes. A server replies
to open-file requests with the server-Id and an identifier of the file or alternatively with
the name of the server managing this file.

The protection scheme is comparable to the one used in UNIX.

Caching in main memory is needed for performance improvements, but is disabled in
the case of concurrent writes, i.e., multiple hosts have a file open at once, with at least
one host having it open for writing. In addition, version numbers are held to change
out-of-date cache blocks when files are sequentially write shared (delayed-write scheme
used). In contrast to NFS, Sprite does not use read-ahead to increase the sequential read
performance.

When a server crashes, the state information is reconstructed using a recovery protocol. On the other hand, if a server does not reply to an open request, the entry in
the prefix table is invalidated and a new broadcast is made. Read-only replication improves availability provided the replicas exist on different servers. The system provides
no atomic updates of replicas. Moreover, updates are not transparent because replicas
have different prefixes.

Miscellaneous

OTHER ISSUES: Processes can migrate to other nodes at any time. Location-dependent
system calls are forwarded back to the home node of a process. If a process has a lot of
dirty pages, the costs for migration are high, but most migrations occur in conjunction
with the *exec* kernel call, and no virtual memory needs to be transferred. Paging can be
done across the network. The amount of main memory used for virtual memory and for
caching of files varies dynamically depending on demand.

STATUS: Sprite is used on a regular basis by a community of 20-25 users at Berkeley and has been distributed to a few dozen sites outside Berkeley. Active.
CONTACT: John Ousterhout, Computer Science Division, Univ. of California, Berkeley, 571 Evans Hall, Berkeley CA 94720, USA
e-mail: spriters@sprite.berkeley.edu
REFERENCES: [232] – [251]

4.21   V

Main Goal

The V distributed system was developed at Stanford University with the purpose of connecting multiple workstations, access and location transparency provided. The small V-kernel provides high performance communication and can be used by servers which offer most other operating system facilities.

Advantages

The main advantage of V is its high-speed inter-process communication which is comparable to local communication.

Description

V is being run on VAXstationIIs, SUN 2/50s and SUN 3s. It consists of a small kernel which has to be booted over an Ethernet from a server on a machine running UNIX 4.x BSD. This way diskless workstations are supported. With a special server UNIX tools can be used, but the UNIX system must be modified in order to support multicasting and an Ethernet filter packet. Not all configurations are supported by the Stanford distribution tape.

V uses a special communication protocol, called Versatile Message Transaction Protocol (VMTP), which supports RPCs and client/server interactions. Messages contain a fixed-length message of 32 bytes and a 32-bit integer that uniquely identifies the destination process. Longer messages can be achieved by sending a pointer to a memory segment, where the first 1 kB of the segment is *piggybacked* onto the message. A client looking for access to a server is blocked until the server has sent back a reply. The protocol provides an at-most-once semantics by using selective retransmissions and duplicate suppression. A special feature is its implementation of multicasting for group communication. All input and output operations use a uniform interface, called UIO, which is block-oriented.

V provides the concept of *process groups* and *teams*. Process groups are any random set of processes that are grouped using kernel support so they can be addressed using a single group identifier, that is part of the same space as process identifiers. One can send to, kill, query, etc., a process group. One can view it as a vast generalization of the UNIX process group mechanism.

A team is a set of processes sharing the same address space. In contrast, it is in fact more usual for members of a process group to be in separate address spaces, for example the group of file servers.

Each process has a unique process-Id. Global unique names are achieved by adding to each object name the name of the server managing this object. If a name is not known a multicast message is sent to all possible servers. To improve performance, (server name, pid) pairs are cached.

There are no special security issues, the same protection scheme is used as in UNIX.

Traditional fault-tolerance, recover techniques, and reliability were not emphasized in the design of V. However, it is worth noting: One reason for the small kernel in V was for reliability. A small kernel is presumably easier to get right, and one can easily replicate server modules outside of the kernel to greater reliability at that level, and this is what V did. For example, multiple file servers provide standard installed files in V, such as system binaries and include files, and the system transparently switches to another server when the one it is using fails, using multicast to locate an alternate. Furthermore, V emphasized the use of multicast. V uses multicast extensively to make the system more fault-tolerant, including decentralizing the naming, scheduling, time synchronization and even experimenting with replicated files and distributed atomic transaction support.

Migration is possible for team address spaces and can be used in a dynamic load-balancing mechanism. To improve the performance of the migration mechanism, the address space of a team is precopied, so that only those pages referenced after this time need to be copied during the period of the real migration.

Miscellaneous

STATUS: V is in use at several universities, research laboratories and companies. Further research is being done in porting V on a multiprocessor machine, connecting it to wide area networks and implementing reliability issues such as atomic transactions and replication.

CONTACT: David R. Cheriton, Distributed Systems Group, Computer Science Dept., Stanford Univ., Stanford CA 94305, USA
e-mail: cheriton@cs.stanford.edu
REFERENCES: [252] – [271]

4.22 WANDA

Main Goal

After some ten years of use of CDCS (Sect. 4.2) a new distributed systems research environment, based on the Cambridge Fast Ring (CFR), a 75Mbit ATM network, was required at the University of Cambridge Computer Laboratory. Software development takes place on Ethernet-based UNIX host systems but a lightweight kernel was needed for the target research environment. A specific requirement was a kernel on which efficient services, including gateways, could be built.

Advantages

WANDA supports multithreaded processes; each process running in an address space and threads sharing an address space. Multiprocessor scheduling is supported. A real time version is being developed which includes periodic threads, dynamic priority and user-specified thread management. The design philosophy is that multiple protocol hierarchies can be supported in a WANDA system. A locally developed Multi-Service (MS) protocol hierarchy is currently installed: MSNL/MSDL. There is a socket-like interface for invoking this service. Memory management is based on paged segments, and page mapping and unmapping is used for efficient interprocess (cross-address-space) communication and for network I/O. Semaphores are supported and are used both in the kernel and for user-level inter-thread communication within an address space.

Description

WANDA runs on VAX-based machines including MicroVAXIIs and the multiprocessor Firefly; on the M 68000 series with and without memory management and on Acorn ARM3s. WANDA machines run on the CFR and on Ethernet and are also used for gateways between the CFR and the 1 Gbit Cambridge Backbone Network which is used to connect CFRs. WANDA is also used in the port controllers of a high-speed space-division ATM switch being developed as part of the UK SERC-supported Fairisle project, in collaboration with HP Research Laboratories, Bristol.

The ANSA testbench, which includes an RPC facility, runs on WANDA and UNIX machines at the laboratories, thus facilitating interworking between the host and target environments. Sun RPC is also installed above WANDA.

An open storage service hierarchy is being developed to run above the WANDA kernel. Prototypes of a low-level storage service (LLSS) and a high-level storage service (HLSS) are operational. The LLSS supports a byte sequence storage abstraction and the HLSS supports structured data, at an acceptable cost, for those clients that require this service. It is intended that clients such as filing systems, database systems and multimedia systems are implemented above these components.

Naming of services conforms to ANSA, naming of storage objects is integrated with a capability interface to the LLSS. Higher-level naming is achieved through Directory Service clients of the HLSS.

Support for transactions is not yet implemented but work in this area is in progress. Non-volatile memory is being used within the storage service for efficiency and reliability.

An authentication service has been developed but has yet to be integrated into the distributed system.

Miscellaneous

The distributed system is intended to support multimedia working, including real-time video. The UK SERC-supported Pandora project, in collaboration with the Olivetti Research Laboratory in Cambridge, supports work in this area including storage and delivery of video and audio and synchronization of the various components of a multimedia document.

STATUS: WANDA is a recent development and the higher level services are still being researched separately and are not yet integrated.

CONTACT: Jean M. Bacon or D. McAuley, Computer Laboratory, Univ. of Cambridge, Cambridge CB2 3QG, UK

e-mail: Jean.Bacon@computer-lab.cambridge.ac.uk

REFERENCES: [272] – [274]

4.23   Wisdom

Main Goal

Wisdom is a distributed operating system for transputers. The objective of the Wisdom project is to make networks of processors easy to use, and their exploitation simple. The approach taken is to provide a scalable operating system, that will run on a scalable architecture, a network of point-to-point connected processors. Wisdom is structured as micro-kernel and a set of basic modules which provide necessary services. The micro-kernel is often viewed as a module in its own right, rather than as a true kernel. Each module is deliberately kept as simple as possible; more complex problems are solve by combining modules.

The system is structured around providing three degrees of transparency (derived from the eight forms described by the ANSA work): parallelism, interconnection and most importantly, location transparencies. Each of these is supported by one of the basic modules of Wisdom: the load balancer, router and namer respectively.

Advantages

Under the studies so far carried out, Wisdom has been shown to be reasonably scalable, and it should be possible to build a system comprising of several hundred processors.

The main disadvantages of this system lie in the current hardware, the absence of memory protection and memory management, and the lack of routing hardware.

Description

The system currently runs on a network of nine T414 transputers which have two disc drives attached (one floppy and one winchester). The network is connected to a SUN host via an RS-232 line. Work is currently underway to port Wisdom to the two Meiko boards

(32 T800 transputers) in the University of York department. The possibility of test runs on the Edinburgh Supercomuter (400+ T800 transputers) is also being investigated.

Wisdom is not a layer, but instead a full operating system, however the design was developed from the work done by several other projects including ANSA, Amoeba (Sect. 5.2), V (Sect. 4.21) and Ninth Edition UNIX.

Communication primitives are send, receive and receive-or-timeout over user process maintained capabilities. The router, which supports these primitives, provides at-most-once communication service using datagrams. Higher-level routines and modules exists which give reliable communication (in the absence of certain types of network failure) and a form of RPC. These do use any form of protocol standard, as they are not connected to any other networks. It is envisaged that network protocols such as Ethernet would be supported by specialist modules which run as ordinary processes.

The namer supports an arbitrary graph (potentially with cycles) to which any process may register itself, and in turn provide more services. Processes find the services they require by contacting the namer to which they were given a connection by their parents.

The load balancer provides process migration only at the point at which the task is created, and not during the processes lifetime. Many researchers have shown that the cost of moving a running task is high compared to balancing a creation request, and that this cost often offsets the advantages gained. It was also impossible to move running tasks on the hardware available.

It has become apparent the routing hardware, such as the TORUS routing chip, are essential for such systems as Wisdom to become scalable. It is also hoped that any such routing chips would be able to deliver messages direct into task's memory (via the memory management unit) without having to interrupt the main processor.

To date no work has been done on the effects of component failure.

Miscellaneous

STATUS: Other work is progressing on the development of a filesystem and caching scheme that could be effectively used on a mesh and that will reduce communication costs and loads. Investigation is also underway into how the parallel provided by Wisdom may be best exploited by programmers. Active.
CONTACT: Kevin A. Murray, Dept. of Computer Science, Univ. of York, York YO1 5DD, UK
e-mail: murray@minster.york.ac.uk
REFERENCES: [275] – [280]

Chapter 5

Object-Oriented Distributed Operating Systems

5.1 Alpha

Main Goal

Alpha is a non-proprietary operating system for providing system and mission level resource management in large, complex, distributed real-time systems. The most demanding examples of such systems are always found first in military contexts (such as combat platform mission management and battle management), but similar needs subsequently arise in aerospace (such as space stations, autonomous exploration vehicles, and air traffic control) and civilian industry (such as factory automation, telecommunications, and OLTP).

Alpha provides location, access, replication, concurrency, and failure transparency.

Advantages

Alpha manages multiple, physically dispersed computing nodes to coherently execute peer-structured transnode computations which cooperatively perform a mission. This global resource management includes transaction-style concurrency control for application-specific correctness of distributed execution and for application-specific consistency of distributed (i.e., replicated, partitioned) data.

Traditional real-time operating systems concepts and techniques have been devised for relatively small, simple, centralized subsystems for sampled-data monitoring and control of primarily periodic physical processes. This has led to a mindset and approach focused on static determinism of both ends and means. However, this technology does not scale up to larger, more complex, more distributed systems because they inherently exhibit extreme uncertainties due to asynchronous, dynamic, and stochastic behavior of both the system and its mission. Alpha manages all physical and logical resources directly with application-specified actual hard and soft time constraints and relative importances for aperiodic as well as periodic results (instead of with priorities). It employs "best effort" resource management algorithms to meet as many as possible of the most important time constraints, given the current mission situation, and taking into account dynamic and stochastic resource dependencies, concurrency, overloads, and faults.

Alpha's disadvantages are principally the logistical and sociological ones which arise from it not having native (as opposed to layered) UNIX compatibility.

Description

Alpha is a native operating system (on the bare hardware) whose kernel is instantiated on every node in a peer relationship ("integrated" not client/server). Alpha's API is unique to Alpha (not UNIX compatible). Alpha's kernel provides a new object-oriented programming model based on "distributed threads" (loci of control point execution) which span objects (passive abstract data types) and (transparently and reliably) physical nodes via operation invocation. Physical location is transparent to thread execution except when the programmer may desire to see or control some of it. This programming model supports coherent distributed (transnode) programming, not just for the applications but also of the operating system itself. Alpha is a multiprocessing operating system; it is fully symmetric, pre-emptive, and multithreaded, even within the kernel.

Alpha is currently operational on 680X0 and MIPS based machines, with ports to others taking place. A multivendor Alpha system of heterogeneous, FDDI-connected multiprocessor nodes will be operational at several user sites in early 1991.

Alpha uses the standard communication protocol XTP (instead of TCP/IP), and adds its own specialized protocols above that. Alpha "IPC" is invocation of object methods, carried out on a synchronous RPC-like protocol, using datagrams.

The semantics of each operation invocation is optionally either at-most-once, or exactly-once using Alpha's transaction mechanisms.

Alpha's naming strategy is object oriented: object methods are accessed via capabilities, which constitute the global name space. There are separate versions of Alpha for discretionary access control and mandatory access control. The standard Alpha does not include encryption, does not employ special hardware components, does use capabilities, and does not include mutual authentication. The B3 multilevel secure version of Alpha (not complete yet) will include encryption.

Alpha is a global operating system in that it incorporates real-time distributed (transnode) resource management policies and facilities in its system layer, built on mechanisms in its kernel layer, to accommodate: asynchronous, real internode concurrency of execution and data access; variable, unknown internode communication delays which prevent global ordering of events; and complex (partial, bursty) failure modes. Thread integrity is maintained by Alpha despite node and communication errors and failures, with at-most-once operation invocation, and orphan detection and elimination, protocols. Transnode, application-specific concurrency control is provided for thread execution ("correctness") and data access ("mutual consistency") using kernel-level mechanisms and system-level policies for real-time transactions.

Availability in Alpha is supported by: application-specific transnode concurrency control based on nested real-time transactions; and replication. Alpha's kernel provides mechanisms which carry out policies that reside above the kernel. A rule-based adaptive fault tolerance strategy manager selects among the policies.

Failure handling/recovery in Alpha is supported by: recovery after crashes; kernel-provided orphan detection and elimination; and stable storage in the kernel.

Alpha's kernel supports dynamic object migration. The policy for assignment of objects to nodes, such as for load balancing and fault tolerance, is above the kernel.

Miscellaneous

Alpha and the distributed applications it supports can be incorporated into a system in three different ways. The cleanest approach to providing a global operating system is for it to be integral when a system is newly designed, in which case there need be only one operating system in the system – native on all nodes – which performs all the local as well as global operating system functionality. But then it is difficult and expensive in performance to accommodate backward compatibility with any extant local system or application software.

Alternatively, Alpha and applications can be added to an existing system by providing an additional set of interconnected hardware nodes specifically for them; these distributed application nodes then interface with the local (subsystem) nodes, operating systems, and applications. This case offers: superior performance due to local/global hardware (node and interconnect) concurrency; compatibility with heterogeneous and pre-existing local subsystems; and relative independence of local and global operating system changes.

A third choice is for the global and local systems to be separate but co-resident on the local nodal hardware – Alpha is able to co-exist with UNIX on any or all of the system's multiprocessor nodes, thus providing the Alpha execution environment for large, complex, distributed real-time applications which are beyond the capabilities of UNIX, together with the UNIX execution environment for relatively small, simple, centralized, real-time applications as well as non-real-time software (for example, window software). STATUS: Alpha initially arose as part of the Archons project (Sect. 8.7) on new paradigms for distributed real-time computing systems at the Computer Science Dept. of Carnegie Mellon University. The prototype was developed there, and demonstrated at General Dynamics, in 1987. The focus of Alpha research and development is now at Concurrent Computer Corp., where a series of next-generation, increasing functionality, versions is being produced. Pilot installations of early releases began at various US Government and industry labs in June 1990.

Alpha is funded by DoD together with a number of companies cooperating on its R&D (including B3 multilevel secure and POSIX-compliant versions, real-time distributed data management facilities, and a real-time rule-based adaptive fault tolerance manager), plus applications for it.

Alpha academic research is taking place at CMU and MIT. Alpha received the highest technical scores in a DoD-sponsored evaluation of operating system interfaces conducted in 1990 by a committee of government, industry, and academic participants.

CONTACT: E. Douglas Jensen, Concurrent Computer Corporation, One Technology
Way, Westford MA 01886, USA
e-mail: jensen@westford.ccur.com
REFERENCES: [281] – [284]

5.2   Amoeba

Main Goal

The main goal of Amoeba was to develop a capability-based, object-based distributed
operating system, which is transparent, reliable and performs well. The idea is to take
a large collection of machines (in any event, much larger than the number of users), and
have them behave like a centralized time-sharing system.

Advantages

The main advantages are its transparency and high performance in a distributed environment. Good use is also made of the available computing power for parallel programming.
Although to the user, Amoeba is UNIX-like, the complete system (kernel, compilers, and
all the utilities) contain no AT&T code whatsoever, so the full source code is available.

The main disadvantages at present are the inability to run existing binary UNIX
programs (programs must be recompiled with a special library), and the lack of virtual
memory.

Description

Amoeba is based on the concept of a microkernel, a piece of code that runs on the
bare hardware of each machine and handles communication, I/O, and low-level memory
and process management. Everything else is done by servers running in user mode. At
present, implementations exist for the SUN 3 and VAXstation. Ports to the 386 and
SPARC are in progress.

An Amoeba system consists of some workstations or X-terminals that provide user
access to the system. Most computing is done on the processor pool, a collection of CPUs
that are dynamically allocated to jobs as needed, and then released (in contrast to the
personal workstation model). For example, a special parallel version of "make" starts up
multiple compilations in parallel on different pool processors to achieve high performance.
In addition, specialized servers such as file servers and directory servers handle certain
system functions, and gateways connect Amoeba systems at distant locations over WANs.
Although many machines are involved, to the user it looks like a single integrated system;
there is no concept of "file transfer" or "remote login".

Logically, the Amoeba system is based on objects (such as files and directories). Each
object has a global unique name, expressed through its capability. A capability contains
the port number of the service which manages the object, the local number of the object
and a field of rights. To keep the port number unique, they are random integers with 48

bits. Capabilities can be stored in directories managed by a directory service allowing hierarchical (UNIX-like) directory trees. To locate the service corresponding to a port, the client's kernel broadcasts a "locate" packet. When the reply comes back, it is cached in the kernel to eliminate the need for subsequent broadcasts. To allow use of services via wide area networks, there is a special server in the local net that impersonates the remote server and forwards all requests to it.

Most of the system's functionality comes from its servers. The bullet server supports immutable files that are stored contiguously on the disk. It achieves a high throughput (800 kbytes/sec). The directory server maps ASCII strings onto capabilities, the same way that UNIX directories map ASCII strings onto inode numbers (in a sense, a capability is like a generalized inode number, except that the object can be anywhere in the world). The run server does load balancing, so when a request to run a process comes in, it is executed on the pool processor with the lightest load. The object server handles replication, to allow the system to maintain multiple copies of files for fault tolerance purposes. Process that need to be fault tolerant can register with the boot server, which polls them periodically and restarts them if they fail to reply correctly. Other servers include a TCP/IP server, an X-windows server, a cron server, a time-of-day server, and various I/O servers.

Miscellaneous

OTHER ISSUES: Amoeba has been used for various applications, including program development and parallel programming. A new programming language, Orca, has been designed that makes parallel programming on Amoeba easier by automatically allowing multiple processes to share user-defined objects in a consistent way. Applications such as the traveling salesman problem, all pairs shortest paths, and successive overrelaxation have been programmed to run on Amoeba taking advantage of the processor pool by starting up numerous processes in parallel.

STATUS: Although Amoeba started out as a typical academic exercise at the Vrije Universiteit, about 90 man-years of work have gone into it so far there at C.W.I. in Amsterdam, and it is now intended as a serious, usable system, with hundreds of pages of documentation and full commercial support available from a company that has been doing UNIX support for 15 years. It will be available (with full source code) by the end of 1990. It has been used as the basis of the MANDIS distributed system (Sect. 8.39) supported by the European Community, and is being used by the European Space Agency as the basis for a system transmitting real-time video over a LAN. Future work will focus on adding support for broadcasting, fault-tolerance, and various other features.

CONTACT: Andy Tanenbaum, Dept. of Mathematics and Computer Science, Vrije Universiteit, De Boelelaan 1081a, 1081 HV Amsterdam, The Netherlands

e-mail: ast@cs.vu.nl

REFERENCES: [285] – [300]

5.3   Argus

Main Goal

Argus is both a programming language for writing distributed applications as well as a system developed to support the implementation and execution of distributed environments. It copes with the problems that arise in distributed programs, such as network partitions and crashes of remote sites. On the other hand Argus does not free a programmer from the concerns of concurrency: deadlocks and starvation are not detected or avoided. Argus is transparent in the sense of access, location and failure.

Advantages

Argus provides two novel linguistic mechanisms: First, *guardians* implemented as a number of procedures that are run in response to remote requests. A guardian survives failures of the node at which it is running (resilient). A guardian encapsulates a resource that is accessed via special procedures called handlers. Second, atomic transactions, or *actions* for short, can be nested and mask problems arising from failures and concurrency.

Argus performs well given that it is a prototype implemented on top of an existing system. Its performance is better than or competitive with that of other systems where performance was much more of an issue in the implementation. Clearly Argus is not ready to be used in a commercial environment since it is only a prototype, but it appears that if someone were to invest the effort to provide a production system, it would have very good performance. Note that calls are done as subactions, which are not expensive. When a non-nested transaction commits, Argus uses two-phase commit, which can be expensive.

Description

The Argus implementation runs on MicroVAX-IIs under Ultrix connected by a 10 Mbit/second Ethernet. It is a new layer on top of Ultrix. No kernel modifications are made except for an optimized UDP, the so-called ACP (Argus Communication Protocol). Argus uses message-based communication (datagram service), and arguments are passed by value. Individual packets are tagged with a message-unique identifier and a sequence number. The final packet indicates the end of a message.

Argus provides built-in types of atomic objects, such as atomic arrays or atomic records and allows a programmer to define his/her own atomic data types.

Every handler call is a subaction and happens exactly once. To send a message, the Argus system first has to map the destination guardian identifier to a network address.

Security issues play no role.

A two-phase commit protocol supports the atomicity of (nested) transactions. Two-phase locking is used to implement the synchronization requirement of serializability. Serializability is implemented via atomic objects providing a set of operations to access and manipulate these atomic objects. Recovery is done using versions written periodically to stable storage. The hierarchical structure of actions (action tree) allows orphan detection by using special state-lists of action participants (orphan destruction algorithm).

Replication of objects is not supported.

Miscellaneous

Argus is also a programming language which gives a programmer the possibility to deal with distributed applications (extension of the CLU language). Argus was unique because it provides atomic actions within a programming language.

Argus has been used to implement a number of distributed programs, such as a library information system, a distributed editor, and a mail repository.

STATUS: The Argus research is active.

CONTACT: Barbara Liskov, Massachusetts Institute of Technology, Laboratory for Computer Science, Cambridge MA 02139, USA

e-mail: liskov@lcs.mit.edu

REFERENCES: [301] – [313]

5.4   BirliX

Main Goal

BirliX is an object-oriented approach providing all levels of transparencies. Its essential design principle is the *abstract data type*. Thereby, its objective is focused on the development of a reliable distributed system. Its architecture is based on an abstract data type kernel/server architecture, in which a small kernel provides the integration of basic services. Higher operating system functions are provided by servers running on top of the kernel. Such an open architecture can be a good basis for the development of reliable distributed systems.

A central part of the BirliX kernel is the abstract data type together with simple inheritance. The basic services of the kernel include definition of abstract data types, their instantiation, their naming (identification), and the communication among instances of BirliX-types. All BirliX-types provide a homogeneous framework for naming, security, fault-tolerance, and distribution issues.

Advantages

The BirliX approach shows that the abstract data type principle does not interfere strongly with performance requirements.

Description

BirliX runs on a network of SUN 3 workstations with a UNIX 4.3 BSD emulation on top of it. The abstract data types as system interfaces were also used in Eden (Sect. 5.9). While Eden is an implementation on top of UNIX, BirliX is a new operating system kernel providing an abstract data type interface with other existing operating systems, currently the UNIX 4.3 BSD. Porting simply consists in relinking the utilities with the BirliX system call library.

There are four fundamental abstractions: abstract data types, teams for implementing abstract data types, and segments, and processes for physical resources.

BirliX uses a uniform and type-independent implementation structure for all abstract data types, the *team*. A team provides all storage and computing resources needed by an abstract data type instance during its lifetime. A team is a collection of threads that are sequential activities running in parallel to each other. The threads share an address space together with a collection of communication bindings to other teams.

Within teams, communication via shared memory is possible and synchronized by monitors. Teams can also communicate synchronously by remote procedure calls in a client/server relationship. Messages that are passed across the network are handled by assisting processes on both sides so that the kernel is kept small (by the way, the next version of the BirliX kernel will handle this message exchange in order to increase the performance).

Error-free message passing is not provided because this can be done more efficiently inside the teams. Supported are timeouts, sequence numbering to detect duplicates at existing client/server relations and searching for previous relations at the start of a new one.

The concepts of its naming services are vital for the survival of any operating system over a longer period of time. The limits of current naming systems become obvious when they are confronted with challenging applications in the field of fault tolerance and security. Flexible naming schemes are needed that provide openness for future requirements.

The performance and reliability of its naming services are vital for the acceptance of a distributed system. Name servers are a suitable example to demonstrate that the BirliX type concept does not interfere strongly with performance requirements.

The BirliX naming model uses concatenable name servers to achieve a high degree of openess. Although many name server properties in the area of fault tolerance and security are very easily achieved in the BirliX type-based environment, the model itself is independent from the actual kernel interface. Its general ideas are also feasible for kernels providing other abstractions.

BirliX is not a capability-based system, such as Amoeba (Sect. 5.2). In BirliX, access to and from an instance is controlled by the kernel. The kernel maintains an access control and a subject restriction list for each instance, defining the access rights of both the using and the used instances.

In the first implementation, users are authenticated using a password-mechanism, but other mechanisms – like the TeleTrusT smartcard – may easily be adapted to the system. To ensure that a proper BirliX system is being used a mechanism called *secure booting* is supported.

Based on the assumption that a single method to achieve fault tolerance is not sufficient for all applications, BirliX provides basic recovery mechanisms to build suitable policies (e.g., transactions, recovery clusters) on top of the system. Since an application program is a set of communicating instances of different types, the system offers efficient mechanisms to (a) checkpoint and recover a single instance in any state, and (b) to re-establish connections between communicating instances.

Hence using a copy-on-modify technique, both for main memory pages and mass storage, checkpointing is supported by the memory management.

Miscellaneous

The notation *team* comes from an analogy with the V-system (Sect. 4.21). In V, a team is a set of processes sharing the same address space; different teams communicate via synchronous message passing.

Threads within a task of MACH (Sect. 4.13) share a common address space; threads of different tasks communicate by message passing through ports. In BirliX, the functionality of a port is achieved by agents.

Another similarity exists between BirliX and Argus (Sect. 5.3). In Argus, abstract data types are implemented by *guardians*, which combine lightweight processes similar to *teams* in BirliX.

STATUS: A prototype of the system has been running since the beginning of 1988. Current work is concerned with the implementation of a new version which includes improved performance, security policies extending the one found in UNIX, simplified internal interfaces, and mobility as well as replication.

It is assumed that the POSIX standard and the functionality of the system interface as defined by X/OPEN can be easily implemented on top of the new BirliX version.

CONTACT: Hermann Härtig, Winfried Kühnhauser, Oliver Kowalski, Gesellschaft für Mathematik und Datenverarbeitung, Schloss Birlinghoven, D-5205 St. Augustin 1, Germany

e-mail: haertig@gmdzi.uucp.dbp.de or haertig@kmx.gmd.dbp.de

REFERENCES: [314] – [319]

5.5 CHORUS

Main Goal

The CHORUS project was launched at INRIA, France, at the end of 1979. The first implementation of a UNIX-compatible micro-kernel-based system was developed between 1984 and 1986 as a research project at INRIA. Among the goals of this project were to explore the feasibility of shifting as much function as possible out of the kernel, and to demonstrate that UNIX could be implemented as a set of modules that did not share memory.

In late 1986 a new version, based on an entirely rewritten CHORUS kernel, was launched at Chorus systèmes. This current version shares most of the goals of the previous and adds some new ones, including real-time support and – not incidentally – commercial viability.

Advantages

CHORUS is a small kernel on which a number of different operating systems can run simultaneously. It tries to specify layered communication protocols unambiguously. This approach, in the early 1980s, has been in line with the work in progress within ISO or CCITT. CHORUS is the first commercial, and the most advanced, micro-kernel-based operating system. Its technology has been applied in the past (CHORUS-V0 and V1) to proprietary operating systems and now to UNIX (SVR3.2, SVR4, BSD 4.3) and also to object oriented environments, such as the ESPRIT projects Comandos (COnstruction and MANagement of Distributed Open System) and ISA.

The main disadvantage at present is the lack of sophisticated fault tolerance.

Description

CHORUS-V3 runs on different computer architectures based on MC 680X0s and Intel 386 (e.g., IBM PC/AT) interconnected through an Ethernet local area network. It provides an extended UNIX System V interface which can be used without changes by existing UNIX applications: the operating systems, for example UNIX, are implemented as servers on top of the kernel. The CHORUS kernel has been ported onto M68020, M68030, M88K, I186, I386, I486, SPARC, VAXes and ARM3 boards or machines; other ports are in progress. The MiX (= UNIX) subsystem runs on a reference machine, today the Compaq 386. On that machine it is binary compatible with SCO SVR3.2. A UNIX subsystem compatible with System V Rel. 3.2 is currently available, with UNIX System V Rel. 4.0 and BSD under development.

CHORUS is a new message passing kernel which structures distributed processing into *actors*. Actors are resource (memory, ports, ...) capsules. *Threads* are executing units; they are encapsulated in actors. Threads communicate and synchronize within an actor through shared memory, or in any other case (i.e., within an actor or between actors), through exchange of messages. Basic protocols for exchanging messages are the asynchronous exchange and the Remote Procedure Call. Threads send and receive messages through *ports*. Whether local or remote, ports are a uniform view of communication. Since ports are mobile, actors do not necessarily know where corresponding ports are located. Port groups provide multicast and broadcast.

Each entity (object, actor, port) is uniquely identified with a global name. This name is not used again for designating any other entity. Objects can be accessed as soon as the actor has a capability of the object.

CHORUS already provides the basic tools for fault tolerance: primary copy replication scheme, port migration, port groups and, soon, atomic broadcast. Other important developments in that area are in progress.

Miscellaneous

OTHER ISSUES: Lightweight RPC, distributed virtual memory and real-time facilities are available. In addition processes can be executed remotely.

STATUS: A prototype implementation was constructed and tested at INRIA to support various distributed functions developed in other research projects such as SIRIUS (Distributed Data Bases) or KAYAK (Office Automation). The research effort was transferred to commercial venture, Chorus système, with CHORUS-V3 as their current version. Interfaces to UNIX SVR3.2 are available. Interfaces to UNIX SVR4 and BSD are in progress.

CONTACT: Marc Guillemont or Michel Gien, Chorus systèmes, 6 Ave. Gustave Eiffel, F-78183, Saint-Quentin-en-Yvelines Cedex, France

e-mail: mgu@chorus.fr (Marc Guillemont) or mg@chorus.fr (Michel Gien)

REFERENCES: [320] – [337]

5.6   Clouds

Main Goal

Clouds is a distributed operating system that integrates a set of nodes into a conceptually centralized system. The system is composed of compute-servers, data servers and user workstations. Clouds is a general purpose operating system, i.e., it is intended to support all types of languages and applications, distributed or not. All applications can view the system as a monolith, but distributed applications may choose to view the system as composed of several separate compute-servers. Each compute-facility has access to all facilities of the entire system.

The goal of Clouds is to provide a simple, integrated distributed system through the use of an appealing system model. Clouds supports a reliable distributed computing environment. The structure of Clouds promotes location, access, concurrency and failure transparency.

Advantages

Clouds provides an integrated distributed environment that can be effectively utilized by both centralized and distributed applications. In addition, distributed applications can be written in a fashion very similar to writing centralized applications.

The advantage of Clouds is its treatment of computation and storage as orthogonal issues using persistent single-level stores (objects) and lightweight processes. The objects can be in use simultaneously at several sites by concurrent processes, making distribution of computation simpler. The provision of sharable persistent stores makes files and messages unnecessary.

Clouds does not provide for inheritance and subclassing at the operating system level. But the operating system provides instantiation and the languages support inheritance for Clouds objects.

Description

The first Clouds version was built on VAX-11/750 hosts which were connected through an 10 Mbit/sec Ethernet. The current version has been implemented for the SUN 3 machines as a native minimal kernel called *Ra* and a set of system objects that provide operating system services. The Clouds environment consists of compute-servers running Ra/Clouds, data servers (currently running on top of UNIX) and user workstations that run UNIX. Clouds provides a distributed resource that is accessible from virtual terminals running under X-windows. Clouds I/O services provide user I/O on these virtual terminals.

Clouds implements an object-thread model of computation that is based on the concepts of object-oriented programming environments. It supports coarse-grained objects that provide long term storage for persistent data and associated code, using a single-level store. Lightweight threads provide support for computational activity through the code in the objects. The object-thread model treats storage and computation as orthogonal concepts.

The object-thread model is implemented at the operating system level. Clouds objects are large-grained encapsulations of code and data that are contained in an entire virtual address space. An object is an instance of a class, and a class is a compiled program module. The objects respond to invocations, and the invocation is the result of a thread entering the object to execute an operation (or method) in the object. Threads are the only active entities in the model.

All objects have unique (flat) system names. User names are provided by directory services.

The Clouds objects are stored in data-servers and the threads execute on compute-servers. The transport of objects between data-servers and compute-servers occur through demand-paging and a memory sharing protocol called Distributed Shared Memory (DSM). This protocol provides access to all objects from all compute-servers and ensures single-copy semantics. Distribution can be achieved by sending invocation requests to any compute-server (RPC-style invocations).

Security issues have not been addressed in the present implementation.

The object-thread model has been extended to provide lockable segments of data within objects and various levels of consistency defined in the operations. This allows Clouds to support a range of computations from atomic actions to process-style update semantics.

Miscellaneous

OTHER ISSUES: Clouds currently supports 2 languages: CC++ (and extension of C++) and C-Eiffel (an extension of Eiffel). These languages provide inheritance fine-grained objects in Clouds objects as well as various forms of inheritance, invocations and memory management. The user interface for Clouds is provided by UNIX.

STATUS: Research in underway to provide programming environments, user interfaces, reliability (in terms of consistency and replication) as well as various topics in distributed systems. Currently, only immutable objects may be replicated.

In the future (when Version 2 is complete) Clouds will be used for experimental research in areas such as distributed computing, networking, databases, blackboard systems, knowledge bases and so on.

CONTACT: Partha Dasgupta, College of Computing, Georgia Institute of Technology, Atlanta GA 30332, USA

e-mail: partha@cc.gatech.edu

REFERENCES: [338] – [354]

5.7 Cronus

Main Goal

The Cronus distributed operating system was developed to establish a distributed architecture for interconnected heterogeneous computers promoting and managing resource sharing among the inter-network nodes. Cronus is based on an abstract object model providing location and access transparency.

Advantages

Cronus is an approach that shows that designing a distributed operating system which operates on and integrates a set of heterogeneous architectures in an inter-network environment is tractable.

The main disadvantage is that Cronus is hardly portable due to the host-dependent kernel implementation.

Description

Cronus is a new operating system kernel which operates in an inter-network environment including both geographically distributed networks with a span of up to thousands of miles, and local area networks. Cronus currently runs on a single cluster defined by one or more LANs. The principal elements of a Cronus cluster include first of all a set of hosts upon which Cronus operates, second, some LANs which support communication among hosts within a cluster, and finally, an internet gateway which connects clusters to a large internet environment.

The current Cronus development uses Ethernet and Pronet Ring LANs. Local network capabilities are indirectly accessed through an interface called *Virtual Local Network* (VLN). It is based on a network-independent IP datagram standard augmented with broadcast and multicast features. The LANs are connected via an ARPA standard gateway (TCP/IP, UDP/IP).

The testbed configuration consists of BBN C70 under UNIX V7, DEC VAXs under VMS and UNIX 4.2 BSD, SUNs under UNIX 4.2 BSD, and GCE 68000 systems.

Cronus is designed as an abstract object model. All system activities are operations on objects organized in classes called *types*. The Cronus IPC is cast in the form of operation invocation and replies, both synchronously and asynchronously. Communication is a superset of the RPC scheme.

Each object is under direct control of a manager process on some host in the Cronus cluster. Associated with all Cronus objects is a unique identifier which is a fixed-length structured bit string guaranteed to be unique over the lifetime of the system. User identity objects, called *principals* (capabilities), are used to support user authentication and the access control mechanisms. The principals are managed by an authentication manager. In a typical login-sequence the user supplies a name of a principal and a password for that principal object.

Some objects can migrate to serve as the basis for reconfiguring the system. Other objects are replicated to increase the availability of the system.

Miscellaneous

OTHER ISSUES: The *Diamond multi-media message system* is an application running under the Cronus operating system.
STATUS: Active. An additional cluster is planned for installation at the Rome Air Development Center.
CONTACT: R.E. Schantz or R.F. Gurwitz, BBN Laboratories Incorporated, 10 Moulton Street, Cambridge MA 02238, USA
REFERENCES: [355] – [357]

5.8   Cosmos

Main Goal

Cosmos has been designed at the University of Lancaster, UK. Cosmos gives integral support for both distribution and programming environment functions. Location and access transparency, and eventually replication and synchronization transparency are provided.

Advantages

Cosmos is not implemented yet, just simulated. However, the Cosmos approach provides a unifying framework for handling distribution and for supporting a more sophisticated programming environment.

Description

Cosmos will be implemented as a new kernel. Each node in Cosmos will run a copy of the kernel. Currently, a simulation of the Cosmos system runs under UNIX. A prototype implementation is planned to run on an Ethernet local area network.

Cosmos is one of the most recent projects that build a distributed system as an object model. An object encapsulates persistent data items and is manipulated by the use of specific operations. All accesses to data objects must be through a well-defined operational interface, similar to Eden (Sect. 5.9) or Clouds (Sect. 5.6).

In Cosmos, objects are *immutable* (see Cedar, Sect. 2.3). Once created, an object will never be changed. Instead, objects are transformed atomically leading to a graph of object versions. To increase the availability of the system, objects may be replicated. An algorithm was developed for maintaining the mutual consistency of the replicas based on Gifford's weighted voting scheme [13].

Miscellaneous

STATUS: Active research on the Cosmos Project stopped in October 1990. The research group has greatly expanded recently (to about 18 researchers and staff) and, partly building on the experiences and lessons from the Cosmos project, is presently investigating a spectrum of O/S issues related to the development of a distributed multimedia applications platform.

A full implementation of the Cosmos system has not been conducted, although the simulation work was developed further to include a partial implementation of a novel transactions scheme capable of supporting nested and long-term transactions.

CONTACT: Gordon S. Blair or John R. Nicol, Department of Computing, University of Lancaster, Bailrigg, Lancaster LA1 4YR, UK

e-mail: cosmos@comp.lancs.ac.uk or john@computing.lancaster.ac.uk

REFERENCES: [358] – [368]

5.9   Eden

Main Goal

The main goal of the Eden project was to combine the distribution and integration of a collection of nodes in a network by use of an object-based style of programming. Therefore it was an important aspect to develop a special language, the Eden Programming language (EPL). It supports remote procedure calls, the use of objects and lightweight processes for synchronous communication. Eden is access, location, replication and failure transparent.

Advantages

The Eden system was implemented in a short time (1 year) and intended to be portable because it resides on top of an existing operating system. This was a conscious choice, in order to shorten development time, knowing it would result in poor performance (although comparable to other contemporary RPC mechanisms). Performance was less important to the main goal, which was to experiment with an object-oriented programming style in several distributed applications. The development of distributed applications is well supported by EPL, including a replicated directory system, a Mail system, and a nested transaction facility.

Description

The final version is operational on a collection of SUN and VAX machines on top of UNIX 4.2 BSD, making it possible to use Eden and UNIX services simultaneously. To run Eden objects, the UNIX kernel was augmented by a run-time library including an IPC package, a lightweight threads package, and a module to compute timestamps.

Objects in Eden mostly communicate synchronously whereby at-most-once semantics is guaranteed. Asynchronous communication was also possible, but experience showed that users always preferred the synchronous mode. Communication was based on the Accent IPC (Sect. 4.1).

Eden objects are referenced through a protected capability which contains a unique identifier and a field with specific rights. Eden objects have an active form (a UNIX process) and possibly a passive form (stored on disk). To locate an object, which can change its location even during an access, caching combined with a special search mechanism is used.

To improve reliability, an object can checkpoint its state to the passive form on disk so that it can be restarted after a crash. The activation is done automatically if the object is invocated and no active form exists. Two replication mechanisms were implemented. One is based on Gifford's weighted voting scheme and replicated only the passive forms [13]. The other used *regeneration* and replicated entire objects. Support for nested transactions has been implemented at the user-level. Among other experiments, the Eden system was used for the first implementation of Bayer's Time Interval Concurrency Control and for distributed load-sharing.

Miscellaneous

Historical remark: EPL was not in the original design. There was a first quick-and-dirty prototype, called "Newark." The Eden programmers complained bitterly about the difficulty of programming. EPL was designed and implemented for the first Eden prototype.

OTHER ISSUES: A garbage collection removes objects without capabilities but this can only work if all objects are passive. To demonstrate the properties and behavior of an object, each has a concrete Edentype which describes all possible invocations.

STATUS: The project is completed and the system was used for a number of experimental applications before being dismantled.

CONTACT: E.D. Lazowska or Jerre Noe, Dept. of Computer Science FR-35, Univ. of Washington, Seattle WA 98195, USA

e-mail: noe@cs.washington.edu

REFERENCES: [369] – [382]

5.10   Gothic

Main Goal

Gothic is a INRIA-BULL research project under investigation at IRISA/INRIA-Rennes. It is a successor of the Enchère system (Sect. 6.6). The objectives of Gothic are to provide a integrated reliable distributed operating system on a network of multiprocessor workstations.

No location or access transparency in Gothic is provided for active entities (processes), only for passive entities. Data can be contained in two kinds of memory units: (a) segments: disk memory units, and (b) stable segments: memory units stored in fast stable storable board, with recovery properties. According to the decomposition of transparency levels, Gothic provides the following transparency properties: location transparency for segments, access transparency for segments, replication transparency for stable segments, synchronization transparency for segments (this scheme provides a system wide consistency at the segment word level) and failure transparency using fast stable storage board to store process checkpoints.

Advantages

Gothic provides a new programming language concept to express parallelism and distribution, a reliable multiprocessor architecture based on stable storage technology, and a segmented virtual memory over a distributed architecture.

Description

Gothic is built on a network of BULL SPS7 multiprocessors (68020 mP, Ethernet communication link).

The system was implemented as a new kernel that runs on every node. Memory management kernel facilities are distributed to all the nodes. Gothic is an integrated system. It investigates a new concept, the *multiprocedure*, which is a generalization of the procedure concepts and nicely integrates procedural control and parallelism. This concept allows, in some ways, a reconciliation between the object-oriented world (based on procedures) and parallelism. Multiprocedures are in particular, well adapted to fragmented-data handling. Unlike traditional data which have a centralized representation, fragmented data may be split up into several pieces, which may eventually be dealt with in parallel. To solve the problem of sharing the secondary memory space between any processes running on the Gothic system, a general virtual memory manager has been designed on a local area network of loosely coupled multiprocessor workstations. This manager provides a single level storage with an invalidation protocol to maintain shared data consistency.

Gothic provides an extension to Remote Procedure Call protocols in order to implement Remote Multiprocedure Call. The protocols rely on some form of atomicity, which is efficiently implemented thanks to the stable storage concept. Gothic uses TCP/IP protocols for intermachine communication. These facilities are not directly used for interprocess communication. Two processes communicate by shared segments. These facilities are used to implement the consistency protocol on segments. Reliable communication protocols were also built, using fast stable storage.

Services on stable segments have an exactly-once semantics. Using stable storages, it is possible to modify a classic multiprocessor machine into a fault-tolerant one: Each processor is equipped with a stable memory board into which it periodically stores process checkpoints. No backup processes are needed since the stable memory associated with each processor can be accessed by another processor through a global bus to retrieve checkpoints after a processor crash. This solution is achievable mainly thanks to the good performance offered by the Gothic version of the stable storage.

Hierarchical naming for segments at the operating system level is provided.

Gothic ensures stable segment integrity (data contained in stable storage is protected against incorrect access from a faulty processor or an errant program). Other security issues are not relevant to the current implementation. Processes can store their state in the Fast Stable Storage Board. Their state is restored after system crashes.

Miscellaneous

STATUS: A first version of Gothic is now completed. It supports network transparency provided by a generalized virtual memory management facility, a reliable subsystem based on the stable storage boards, a parallel object-oriented programming language implemented on the distributed architecture.

The next step of the project is to redesign the parallel object oriented language in order to integrate exception handling mechanisms and to provide an appropriate solution to the difficult problem of making inheritance and synchronization fit together. Second, a new version of object run-time support on a multiprocessor architecture will be designed, addressing issues like protection, load-balancing, distributed garbage collection, and tranparency of fault-tolerant mechanisms at the object programming level.

CONTACT: Michel Banatre, INRIA, IRISA, Campus de Beaulieu, F-35042 - Rennes Cedex, France

e-mail: banatre@irisa.fr

REFERENCES: [383] – [391]

5.11   Grenoble Universities Integrated Distributed Environment (Guide)

Main Goal

The Grenoble Universities Integrated Distributed Environment (Guide) is an object-oriented distributed operating system for local area networks. It embodies the object-oriented architecture defined in the ESPRIT Comandos project (COnstruction and MANagement of Distributed Open Systems). The main design decisions comprise first the use of an object model which includes the notion of separate type and classes, with simple inheritance and second, the dynamic migration of multi-threaded jobs among network nodes. The object model is also supported by a language for developing distributed applications, with emphasis on cooperative processing (for example document management) and interaction.

Advantages

Guide provides location transparency of network resources to the application layer. It is a convenient tool for developing transparently distributed applications that use persistent data.

Guide is still a research prototype, with the deficiencies that may be expected at this stage (lack of support and robustness).

Description

Guide is designed as a set of distributed services, which may be viewed as the upper layers of a distributed operating system. Its current version (1.5) runs on top of UNIX, on the following machines: SUN 3, SUN 4, SUN 386i, Bull DPX1000 and DPX-2, DEC-station 3100.

Future versions are expected to be developed on a low-level kernel.

Guide uses persistent objects, i.e., the object's lifetime is not related to the execution of programs, but the object is in existence as long as at least one other persistent object refers to it. Each object is named by a system-wide unique identifier. Objects encapsulate a set of data and a set of procedures. A *type* describes the behavior shared by all objects of that type. A *class* defines a specific implementation of a type. Classes are used to generate instances. To define a specialization of a given type, subtyping can be done. The hierarchy of subtypes among types is paralleled by the subclassification of classes. A subclass inherits the operations of its superclass, and can overload them. Generic types and classes are also provided. Objects are only accessible through typed variables which are used to invoke operations on objects.

The computational model is based on jobs and activities. A *job* is a multiprocessor virtual machine which provides an arbitrary number of sequential threads of control, called *activities*. Jobs and activities are transparently distributed. Activities communicate through shared objects, which include a synchronization mechanism.

A nested transaction concept is implemented which provides exactly-once semantics to atomic objects. The property of an object to be atomic is fixed at the time the object is created. Each transaction is entirely contained within a single activity.

Failure handling and security issues are not discussed due to the current status of the project.

Miscellaneous

OTHER ISSUES: The Guide language supports the object model described above. It is a strongly typed language with conformity, single inheritance and persistent objects. There is support for synchronization, genericity and exceptions.

STATUS: Guide is a joint project of Bull and the IMAG Research Institute of the Universities of Grenoble. It is conducted in the Bull-IMAG joint research laboratory. Both organizations are member of the ESPRIT Comandos project. Version 1.5 of Guide currently runs on UNIX. Version 2 is under design and is expected to run on MACH (cf. section 4.13).

CONTACT: Sacha Krakowiak, Bull-IMAG, 2 Ave. de Vignate, F-38610 Gieres, France

e-mail: krakowiak@imag.fr

REFERENCES: [392] – [400]

5.12   Gutenberg

Main Goal

Gutenberg is an operating system kernel designed to facilitate the design and structuring of distributed systems and to provide all forms of transparency. The crux of the Gutenberg approach is its use of port-based communication, non-uniform object-orientation and decentralized access authorization using ports and the *Interconnection Schema*, a repository of information about potential process interconnections.

Advantages

Problems of performance within systems that support protection should be solved in Gutenberg by structuring objects at the operating system level only if they are shared. Objects are assumed to be large to amortize the cost of communication.

Further experimentation with the kernel will eventually determine the feasibility of this approach.

Description

Processes need ports to communicate with each other. There are three types of ports (send, receive, send/receive). Messages are sent synchronously (with acknowledgement), asynchronously or RPC-like (via TCP/IP). Interprocess interactions occur under exactly-once semantics.

The creation and use of ports are controlled by the use of capabilities. All capabilities that grant access to potentially sharable objects are listed in the Interconnection Schema (IS) and are further organized into groups, called *IS-nodes*. IS-nodes are identified by system-wide unique names for the kernel's use and by user-specified names. IS-nodes are linked to other IS-nodes via capabilities. Capabilities in IS-nodes are stable and sharable. Each process has an associated IS-node and a set of transient capabilities, both of which are owned by this process. The transient capabilities are destroyed when the process terminates.

Each creation of and access to user-defined objects involves the checking of kernel-defined capabilities. Therefore, authorization is guaranteed if each kernel is trusted. Capabilities can be transferred permanently or temporarily. Each transfer is monitored by the kernel. Capability transfers can be used in conjunction with object migration and for load balancing purposes.

The interconnection schema is partitioned and replicated on each site. IS-nodes are replicated in locations at which they might be used. The site at which an IS-node is created keeps track of the sites with copies of the same IS-node.

Synchronization of access as well as consistency of the replicas are preserved in a *two-phase-locking scheme*, either with a primary site or by a distinction between read and write. Individual object managers can control concurrent access to the objects using different schemes including those that exploit the semantics of the operations defined on the objects. Gutenberg provides a means to decompose the execution of processes into units of computation (called *recoverable communicating actions*) that execute atomically with respect to kernel objects and stable storage. Communication leads to dynamically developed dependencies between these actions. Stable storage and a two-phase commit protocol are used to ensure atomicity.

Miscellaneous

STATUS: Prototypes of the Gutenberg kernel have been built over a number of operating system platforms: A distributed version has been implemented on a network of VAXstations running Ultrix. A multiprocessor version has been implemented on the Sequent Symmetry multiprocessor running Dynix. Each implementation was tailored to exploit the underlying operating system and hardware architecture. No changes had to be made to either Ultrix or Dynix to carry out the implementation. An optimized version might require changes to the underlying operating system. The two implementations have shown that Gutenberg can serve as the unifying layer that lies on top of a native operating system to construct a distributed operating system based on Gutenberg principles. Gutenberg principles have also been validated by its use in structuring a distributed office system.

CONTACT: Krithi Ramamritham, Dept. of Computer and Information Science, Graduate Research Center, Univ. of Massachusetts, Amherst MA 01003, USA
e-mail: krithi@cs.umass.edu

REFERENCES: [401] – [412]

5.13   MARUTI

Main Goal

The MARUTI real-time operating system is being developed at the University of Maryland at College Park. It is designed to support hard real-time applications on a variety of distributed systems while providing a fault tolerant operation. It is an object-oriented design and provides a communication and an allocation mechanism that allows transparent use (location and access) of the resources of a distributed system.

Advantages

MARUTI supports guaranteed-service scheduling, in which jobs that are accepted by the system are guaranteed to meet the time constraints of the computation requests under the requested degree of fault tolerance. As a consequence, MARUTI applications can be executed in a predictable, deterministic fashion.

It is just implemented as a proof of the concept. Thus, system objects run as user processes that are subject to all UNIX idiosyncrasies. There are no resource manipulators in the implemented prototype, so timing properties cannot be verified, and performance measurements cannot be evaluated. Due to the inherently difficult problem of scheduling and allocation, it is not better than other systems in pre-allocation time or resource consumption.

Description

As a new layer, MARUTI runs on top of any UNIX platform, since there are no UNIX primitives used to perform its functions that are unique to a single system. It currently runs on SUN 3, DEC 3100, SPARC and VAXstations.

In this prototype, UNIX facilities are used for message passing in a lightly loaded Ethernet. MARUTI is implemented with one-way invocations, not allowing RPCs. Exactly-once semantics comes from the dispatcher and the resource allocation scheme: everything is reserved for exactly one instance of the service to execute (of course, for periodic jobs it depends on the number of periods).

MARUTI is organized in three distinct levels, namely the kernel, the supervisor, and the application levels. The kernel is a collection of core-resident server objects. The kernel is the minimum set of servers needed at execution time and is comprised of resource manipulators. It is an object-oriented system whose unit entity is an object. While the concepts of object and the encapsulation have been used in many systems, in order to incorporate these concepts in a hard real-time system some extensions have been made in MARUTI. Objects in MARUTI consist of two main parts: a control part (or *joint*), and a set of service access points (*SAPs*), which are entry points for the services offered by an object. A joint is an auxiliary data structure associated with every object. Each joint maintains information about the object (e.g., computation time, protection and security information) and its requirements (service and resource requirements). Timing information, also maintained in the joint, is dynamic and includes all the temporal relations among objects. Such information is kept in a calendar, a data structure, ordered by time, which contain the name of the services that will be executed and the timing information for each execution.

Objects communicate with one another via semantic links. These links are called semantic because they also perform type and range checking in the values they carry. This concept permits implementation of semantic checks as a part of the implementation of the communication link. Objects that reside in different sites need agents as their representative on a remote site. The link is responsible not only for the remote transmission of messages but also for the external data translation of these messages. Security functions can also be specified for the semantic links, such as decryption and encryption. The object principle and the use of the joints allow each access to an object to be direct, and the binding philosophy of the operating system supports it uniformly. Access to an executing object is an invocation of a particular service of that object. For timing reasons, all access rights are checked at pre-runtime.

MARUTI views the distributed resources as organized in various name domains. A name domain contains a mutually exclusive set of resources. This concept is used in the implementation of the fault tolerance using replication. In addition, such division of resources is useful for the distribution, load balancing, faults independence, and feasibility of fault tolerant schemes. The use of joints, and specifically of calendars, allows verification of schedulability, reservation of guaranteed services, and synchronization. Projection mechanisms support propagation of time constraints between different localities. These projections maintain the required event ordering and assure the satisfaction of the timing constraints. Furthermore, these data structures facilitate the management of mutual exclusion and protection mechanisms. Exceptions and validity tests are reduced after a semantic link is established. The links are created by the binding and loading processes and the protection mechanisms are activated and authorizations are established prior to run-time, to allow direct access afterwards. Semantic links to remote objects are established through their respective agents.

A capability-based approach is used for protection and security. This system is completely predefined prior to execution of the jobs. The necessary information for the capabilities are stored in the joint, and the capability itself is furnished by the user.

Fault tolerance is an integral part of the design of MARUTI. The joint of each object may implement fault detection, monitoring, local recovery, and reporting. Initially, each joint contains a consistency control mechanism to manage alternatives (redundant objects with state information) or replicas (redundant stateless objects). The resource allocation algorithm supports a user-defined level of fault tolerance, where redundancy can be established temporally (execute again) or physically (parallel execution of alternatives).

Miscellaneous

The development of hard real-time applications requires that the analyst estimate upper-bounds for the resource requirements for all parts of the computation and that the system ensure availability of resources in a timely manner to meet the timing constraints. Since this is a very cumbersome process, as a part of the MARUTI system, a set of tools have been developed to support the hard real-time applications during various phases of their life cycle.

There are two types of jobs in MARUTI, namely real-time and non-real-time, justifiable due to the requirement of a reactive system (i.e., those that accept new jobs while executing already-accepted guaranteed jobs). A real-time job is assumed to have a hard deadline and an earliest start time. For non-real-time jobs, no time constraints are specified and, therefore, jobs are executed on the basis of time and resource availability. Priorities can be easily incorporated, for example, by implementing a scheme for the revocation of jobs or a multi-priority queue.

STATUS: The current version is a prototype, running on several UNIX platforms, and is intended only for the proof of concept, not as a full system. It is running at the University of Maryland, College Park. This version supports a distributed environment, with a user-defined level of fault tolerance. The prototype runs using a virtual clock, and is dependent on the underlying system calls. The system is currently being ported to operate on top of MACH (cf. section 4.13), due to the similarity in primitives offered. CONTACT: Olafur Gudmundsson or Daniel Mosse', Dept. of Computer Science, University of Maryland, College Park MD 20742, USA
e-mail: ogud@cs.umd.edu or mosse@cs.umd.edu
REFERENCES: [413] – [418]

5.14   NEXUS

Main Goal

NEXUS is a distributed operating system designed and implemented to support experimental research in object-oriented distributed computing and fault-tolerance techniques. This system has been implemented on a network of SUN workstations with some extensions to the UNIX 4.2 kernel. It provides an object-based computing environment and supports network transparency in accessing objects.

Advantages

NEXUS provides the flexibility of implementing tailored object manager designs for experimentations, which can include use of different recovery and concurrency control mechanisms. The standard set of libraries provided for building object managers support nested transactions based on two-phase locking and two-phase commit.

One of the advantages of the NEXUS system is that all UNIX functions are completely visible at the application level.

The primary disadvantage of the current system is that it takes a few minutes (about 10–20) to compile and install a new type definition in the system, and the nested transaction system tends to degrade the system performance.

Description

The system consists of a minimal distributed kernel, which provides inter-object communication, and a collection of system-defined objects, which provide the operating system services. The NEXUS kernel has been integrating with the UNIX 4.2 kernel on SUN workstations and about 25 new system calls have been added to the UNIX system call library.

The NEXUS kernel provides asynchronous invocation between objects. Each object is assigned a 64-bit UID, which is used by the kernel for supporting inter-object communication. The kernel participates in the object search and location algorithm in conjunction with other system objects. Invocation messages which are short (1 kB) are directly transferred from one kernel to another using User Datagram Protocol (UDP). Long messages are communicated across hosts using user-level processes, which also communicate using UDP.

Each object is controlled by an *object manager* that schedules processes to execute operations on that object. All objects of the same type are grouped into a *class*. UNIX processes managing these classes are called *class representatives*. Each representative manages a subset of the objects of its class and maintains a registry of all objects of its class with their most probable location in the network. The kernel maintains a cache which is used to locate the representative managing an object; the cache entries are periodically "aged" and at some point deleted. New cache entries are added whenever an invocation is successful.

A set of standard library functions for transaction management and concurrency control are provided. All invocations within a transaction are treated as nested transactions. NEXUS introduces the concept of *weak atomicity* meaning that atomicity of an action on an object is guaranteed only for changes made to that object's state.

The RPC system is implemented above the kernel, at the object manager level. The standard RPC library provides at-most-once semantics of the remote procedure execution. At the client-side, on time-out, retransmissions of the request message take place for some number of times, and this number can be changed dynamically. Any duplicate invocation from the same client is detected by the server-side object manager. If exactly-once semantics is desired, the transaction primitives must be used. This RPC system supports invocation of asynchronous calls and synchronization with the return of any of the pending asynchronous calls. The synchronization primitives update the "out" parameters of only those calls which return without any error. Some of the interesting features of this RPC system include support for dynamic binding, type checking parameters at the server level, object-oriented and modular programming by supporting dictionaries of interface methods, and multi-level RPC which results in peer-to-peer calls at multiple layers during a single RPC invocation.

Two system-defined objects, *NexusType* and *TypeDef*, support creation and management of new object types in the system. Another system-defined object called *Nexus-Name* supports use of symbolic names and maps such names to UIDs. These objects are replicated to provide high availability of the operating system services. An interactive window-based command shell provides interactive access to the NEXUS environment.

Security does not play any role.

Miscellaneous

OTHER ISSUES: A high level language called *Onyx* has been implemented on the NEXUS system for supporting object-oriented distributed programming. This language has Pascal-like syntax and supports data abstraction and multiple inheritance. It also provides high-level constructs for parallel invocation of operations and recovery from errors. The standard nested transaction management functions are also available to the programmer for building reliable applications.

STATUS: A prototype is currently operational on a network of SUN 3/50 and 3/60 workstations. It is being used in many experimental studies such as performance evaluation of reliable directory update protocols, implementation of reliable broadcast protocols, building fault-tolerant application systems, and design of fault-tolerant primitives for parallel programming using Onyx.

CONTACT: Anand Tripathi, Dept. of Computer Science, Univ. of Minnesota, Minneapolis MN 55455, USA

e-mail: tripathi@cs.umn.edu

REFERENCES: [419] – [423]

5.15   Process Execution And Communication Environment (PEACE)

Main Goal

The main goal of PEACE is to provide a high-performance distributed Process Execution And Communication Environment for massively parallel systems. PEACE is dynamically alterable and application-oriented, thus offering an open operating system interface.

PEACE is an object-oriented distributed operating system providing location, access and replication transparency.

Advantages

PEACE provides a very fast inter-process communication, network transparency, and a highly modular (object-based/process-structured) system organization. It is a highly distributed (decentralized), application-oriented system structure.

At present, PEACE provides no multi-user security, has a poor programming environment and limited I/O capabilities.

Description

PEACE runs on SUPRENUM hardware and other 680X0-based systems. It is a new "from scratch" implementation, structured into: *nucleus*, supporting system-wide communication and threading, *kernel*, providing scaling transparency and hardware-abstractions, and *server*, which is application-oriented and completely executed in user-mode. It exists as a UNIX "guest layer" implementation.

UDP is used for UNIX access, and a problem-oriented (blast) protocol is used for the SUPRENUM network. PEACE offers an RPC facility, including stub generators for Modula-2 and C. All kernel and server services are RPC-based; only nucleus services are invoked by traps. Multithreaded server are supported.

The services provide may-be, at-least-once and exactly-once semantics. What semantics will be used depends on the type of service and on application requirements.

The naming scheme is object-based, hierarchical and application-oriented (domains). The name space is implemented by distributed name servers and local name caches. Its structure is dynamic alterable.

PEACE provides no security issues yet. Availability is increased by means of replicated servers. Failure handling, process migration, or mobile objects are not yet supported.

Miscellaneous

PEACE is object-based and provides a framework for a family of operating systems. It has exchangeable (low-level) system components and shows for an open operating system interface. Besides, PEACE is configurable by a third party.

STATUS: A stable version for SUPRENUM is available now. Future work will focus on virtually shared memory, fault tolerance, incremental loading and auto-configuration, process migration, security and UNIX emulation.

CONTACT: Wolfgang Schröder-Preikschat or Friedrich Schön, GMD FIRST, Technische Universität Berlin, Hardenbergplatz 2, D-1000 Berlin 12, Germany
e-mail: {mosch,fs}@gmdtub.uucp.dbp.de
REFERENCES: [424] – [429]

5.16 Profemo

Main Goal

The main goal of the Profemo design is to provide nested transactions in a distributed object-oriented environment.   Profemo guarantees location, access and concurrency transparency.

Advantages

Profemo is an object-oriented approach resulting usually in poor performance. To compensate, inheritance is not defined on the language level, but is an integral feature of the Profemo design and supported by hardware.

Description

Profemo stands for "Design and Implementation of a fault tolerant multicomputer system" developed by the *Gesellschaft für Mathematik und Datenverarbeitung* in Germany. Profemo is a new layer on top of UNIX 4.2 BSD. Profemo's architecture is divided into subsystems which communicate through messages via stream-sockets.

Profemo is based on an object model. Each object consists of three distinct parts, the object header containing semantic information about the object in order to achieve a well-defined interface, the capability part exclusively holding references to other objects, and the data part.

For some of the extended objects, a so-called *invoke* function is defined that executes a type-specific operation providing simple synchronous calls and concurrent calls. Simple synchronous calls are RPC-based and use the Unreliable Datagram Protocol (UDP). Parameter passing is supplied by a parameter object which allows call by reference and call by value. Concurrent calls correspond to a traditional *fork*. As a result, the calling process is blocked until all results have been delivered indicating that all called processes have been terminated. Each parallel path of computation is a simple synchronous call in itself.

Transactions are the main form of computation. Communication between transactions is performed via shared data objects. A lock descriptor holds the information necessary for the non-strict two-phase locking protocol which guarantees the serializability of the operations performed on a shared object.

Capabilities are used for access control. Further sophisticated security issues have not been implemented.

As mentioned above, concurrency control is achieved by locking within a transaction scheme. Transactions can be nested. The transaction manager had a great influence on the BirliX recover design (Sect. 5.4).

Miscellaneous

OTHER ISSUES: *Mutabor* (mapping unit for the access by object references) represents an efficient support meant to tailor a specific memory management unit. Mutabor provides efficient save/restore facilities and basic functions to manage the recovery points needed by the transaction scheme.

STATUS: Research is going on to look for alternative synchronization policies and to increase the system's performance.

CONTACT: E. Nett, Institut für Systemtechnik der Gesellschaft für Mathematik und Datenverarbeitung mbH, GMD, Schloss Birlinghoven, D-5205 St. Augustin 1, Germany
e-mail: nett@gmdzi.uucp.dbp.de

REFERENCES: [430] – [435]

5.17   Prospero

Main Goal

Prospero is a distributed operating system based on the Virtual System Model. The Virtual System Model is an approach to organizing large systems that makes it easier for users to keep track of and organize the information and services that are of interest. Access to information and services in Prospero is location and access transparent. Replication transparency is planned for the future.

Advantages

Prospero allows users to build their own "virtual systems" from the files and services that are available to them. It is easier for users to keep track of objects in the resulting name space. Tools are provided that allow a name space to be specified as a function of other name spaces. This relieves part of the burden of keeping name spaces up-to-date.

A disadvantage of supporting multiple name spaces is the lack of name transparency: the same name may refer different objects when used by different people. Prospero addresses this problem by supporting closure.

Description

Prospero is presently implemented for the UNIX operating system. Implementations are planned for other systems. Implementation will be interoperable. Objects that are accessible through Prospero may reside on systems of different types. The Prospero protocol makes extensive use of type fields. This allows full support for heterogeneity. Because Prospero is intended to scale across administrative boundaries, support for heterogeneity was an important requirement in its design.

Prospero runs as a network service. Each system that contains objects that are to be accessed through Prospero must run the server. Client applications can run on any system, not just those running the server. No kernel changes are required. Several applications supporting the Prospero name space are provided and a library redefining the open system call allows existing applications to use the Prospero name space. A kernel implementation is planned. Such an implementation would avoid the need to relink applications, and would thus reduce the size of the resulting application binaries. The kernel and non-kernel implementations will interoperate, so users can choose whichever implementation is most suitable for their environment.

The Prospero protocol is presently implemented on top of UDP (in the TCP/IP suite of protocols). The protocol could be easily implemented on top of other protocols, and simultaneous support for multiple protocols has been made possible by the extensive use of types, especially in passing network hostnames and addresses.

A *virtual system* defines a view of the world centered around the user. Those files and services of interest to the user have short names, while the names of objects that the user is less likely to access are much longer. Users can specify parts of their name space as a function of one or more other name spaces. This is accomplished using two tools: a filter is a user defined program, associated with a link, which changes the way one views objects seen through that link; a union link takes the contents of a linked subdirectory and makes them appear as if they are part of the directory containing the link.

Naming in Prospero is "object-centered". In object-centered naming, multiple name spaces are supported, and each object has an associated name space (forming a closure). By supporting closure, it is always clear which name space it to be used when resolving a name. To the user, the name space appears hierarchical. The global name space, however, forms a generalized directed graph. A virtual system specifies the node in the directed graph that is to be the root of its associated name space. When projected from that root, the name space appears to be hierarchical. The directed graph that forms the global name space is said to be generalized because of support for filters and union links.

Access to individual objects in Prospero is mediated by existing access mechanisms, and the security for such access is defined by those mechanisms. Access to read or write directory information may be specified by access control lists associated with directories, or with individual links within a directory. Prospero uses interchangeable authentication mechanisms. Authentication information is passed as an authentication type field, and a field for data specific to that type. The prototype uses an authentication type of "assertion", a future release will incorporate Kerberos authentication (Sect. 6.1), and other authentication mechanisms may be added as needed.

Objects in Prospero may move around, and the links to the object will continue to work. This is accomplished through a combination of forwarding pointers, callbacks, and timeouts.

Miscellaneous

STATUS: A prototype is running at the University of Washington and at a few sites elsewhere on the Internet. A beta release will be available shortly.
CONTACT: Clifford Neuman, Dept. of Computer Science and Engineering, Univ. of Washington, Seattle WA 98195, USA
e-mail: bcn@cs.washington.edu
REFERENCES: [436] – [439]

5.18   SOMIW Operating System (SOS)

Main Goal

The SOMIW Operating System (SOS) is an experimental distributed object-oriented operating system. SOMIW, an ESPRIT project, stands for Secure Open Multimedia Integrated Workstation.

The main goal of the SOR (French acronym for Distributed Object-Oriented Systems) group of INRIA is to implement an object-oriented distributed system which offers an object management support layer common to all applications and languages. This should offer a more simple universe for the development of applications, facilitate the implementation of object-oriented language compilers, make applications more efficient, and allow independent applications to communicate and share objects, without prior arrangement. Each application should only pay the price of those mechanisms it really needs and uses.

Advantages

Due to its general approach, SOS is a positive experience. It is useful for many different applications. The main disadvantage at present is the lack of preemptive task scheduling, and concurrency control mechanisms.

Description

A prototype of the SOS kernel and system services have been implemented in C++, on top of UNIX (SunOS). It provides a set of replaceable protocol objects built on top of a flexible transport protocol. In this way a reliable protocol can be chosen which detects and recovers lost messages, crashes and network failures and suppresses orphans. On the other hand, an unreliable protocol can be chosen as well. The maximum size of messages is given by the size of the network packet. On each host, there is one protocol manager for each type of protocol. On request it allocates a protocol object, connects it to the caller and to a protocol object on the callee's host. This virtual connection is maintained indefinitely, or separately released by the protocol manager should the resource be needed. Normally, an RPC is used for client/server communication but message passing (including multicasting) is possible as well.

SOS is based on a concept of "distributed object", implemented as a "group" of elementary objects distributed among different address spaces; members of a group have mutual communication privileges, which are denied to non-members. Access to a service may occur only via a local *proxy*, which is a member of the group implementing the service. Typically, a client gains access to some new service by importing (migrating) a proxy for that service. Each object is designated by its address within a context (separate address space provided by the SOS-kernel), or globally by a reference. The reference contains a unique identifier, called the concrete OID, and a location hint. An extension to the object concept is implemented in terms of distributed or *fragmented objects*. Fragmented objects are groups of elementary objects that are allowed to be located in several contexts, i.e. on different sites. Thereby, the fragments of a fragmented object communicate via cross-context invocation, or shared memory.

The system services are structured as fragmented objects. Four basic system services are available: The *acquaintance service* is the distributed object manager. It deals with localization, migration of objects, in cooperation with the communication and storage services. Each context has an acquaintance service proxy at instantiation which yields to the context the basic operations on elementary objects. The *communication service* provides communication between sites, and a set of invocation protocols allowing remote procedure calls and multicast. The *storage service* handles the generic aspects of object persistence. It defines a minimal set of simple and generic tools to make storage of typed composite objects handled quasi-automatically by the system. The *name service* manages the binding of internal names to symbolic names. Any object can be named in the same manner. The name service allows clients to build their own view of the name space.

Naming in SOS is based on a two-level hierarchy: the first level is supported by references, which is very user-unfriendly. Therefore, a second level provides symbolic names. A name service maps the symbolic names to the internal references.

Access to the internals of a fragmented object is poorly protected. This was considered acceptable for a proof-of-concept prototype, but must be fixed for SOS to evolve into a real system. Protocol managers recover from a crash by sending a multicast to the other hosts to retrieve the lost state. Crashes of hosts are detected by probe messages which are sent out periodically. SOS supports object migration, persistent objects, dynamic linking, and arbitrarily complex user-defined objects (written in C++).

Miscellaneous

SOS is thoroughly documented with a reference manual (similar to the UNIX manual) and an introductory programmer's manual. Most of the code is in the public domain, except for a few components derived from AT&T code, for which an AT&T licence (UNIX and C++) is necessary.

STATUS: A prototype, built from 1985 to 1988, is currently being distributed to the partners of the ESPRIT project. It has been used as a testbed for the SOMIW applications, such as BFIR2, a multimedia document tool-box, as well as Images, a user-interface management system.

CONTACT: Marc Shapiro, INRIA, B.P. 105, F-78153 Rocquencourt Cedex, France
e-mail: shapiro@sor.inria.fr or sos@sor.inria.fr

REFERENCES: [440] – [452]

Chapter 6

Closely Related Systems

6.1   Athena

Main Goal

Athena is a joint project of the Massachusetts Institute of Technology (MIT), DEC and IBM to explore the potential uses of advanced computer technology in the university curriculum.

In general, Athena is a distributed workstation system. It is a single unified model, similar to a large timesharing system; it makes use of the network services model of computing to replace several functions which are ordinarily available in the timesharing model of computing. Athena provides to its users some sort of location and access transparency through dedicated service machines.

Advantages

In the beginning, Project Athena was an educational experiment introduced in May 1983. Today, Athena is a fully operational production system consisting of 1000 workstations in 40 public clusters and many staff and faculty offices, for use by students, faculty and staff 24 hours a day.

Athena is a novel approach because it is the first attempt to use and integrate the computer power of a large campus-wide network into an academic curriculum.

Description

MIT has installed a network of about 1000 high-performance workstations. The network is implemented with multi-vendor technologies (such as the Proteon Ring) and is based on a high-speed backbone network. Existing campus facilities have access to subnets connected to this backbone network via special purpose gateways called "C gateways." Typically, Athena workstations are dataless nodes.

File service is provided over the network with three distinct protocols. Users usually need not know which protocol provides their personal file service, since these details are abstracted behind a model where all files live in *lockers*. Lockers of all types are accessed with the "attach" command, which does the appropriate lookups and performs the appropriate service requests to make the files in a particular locker accessible to the user.

Remote Virtual Disk (RVD) is used to provide read-only file service; the most common application for RVD service in the Athena environment is the provision of system software. The RVD protocol was created at MIT's Laboratory for Computer Science, and is not available as a commercial product; Athena is moving away from this type of file service in favor of more modern ways to provide networked file service which are better-suited to large distributed environments.

SUN NFS (Sect. 2.8) is commonly used in the Athena environment to deliver files which can be changed by the user; this has typically been used to provide file service for users' home directories, as well as filespace for cooperative work.

The Andrew File System (AFS) (Sect. 2.2) holds read/write files as well. However, AFS is especially well-suited for use in distributed environments and for files used for cooperative work.

Athena used in the past a single operating system and a single communication protocol family to implement its computing environment. The operating system is UNIX, the protocol family is TCP/IP. This transport protocol is used to design the same virtual network for all machines in the project. Use of the TCP/IP protocols continues, but network clients for the Athena network services are in the works for PC-DOS and Macintosh Operating System users.

User-Id and login names are unique across all Athena nodes. The name space of files is UNIX-like and hierarchical.

Computers connected to the network can be divided into two categories: service computers and workstations. Service computers provide generic services such as filing or authentication to one or more client computers. It is possible to share information and to access data and programs from any computer. Servers provide this access to system libraries, to class-specific libraries, and to user lockers.

Athena users are able to "log in" once at the start of each session. The login process authenticates each user, which in turn enables them to use network services as needed. Authentication is accomplished using *Kerberos*, an authentication service developed at Project Athena. Kerberos is an private-key-based authentication service which uses a modified version of the Needham and Schroeder key distribution protocol. Kerberos has since gained acceptance beyond Project Athena, and has been incorporated with a number of other systems. Kerberos is used in the Andrew File System. It has also been adopted by OSF, and SUN is in the process of incorporating it with their remote procedure call.

Global replication schemes, failure transparency and network-wide location transparency are not completely provided; some services operate seamlessly through failures. Athena does not support process migration.

Miscellaneous

STATUS: Athena is still in use for student education. Athena's two industrial sponsors have committed funding through June 1991. In the remainder of the Athena project, the following aspects are covered: move to private ownership of the workstations, shift to supported and automatically updated software, provide dynamic network reconfiguration, support PCs, and deploy to other universities. At present, about ten installations of Athena exist throughout the world. Others are being considered.

CONTACT: Janet Daly, Information Officer, MIT Project Athena, E40-358F MIT, One Amherst Street, Cambridge MA 02139, USA

e-mail: info-athena@athena.mit.edu

REFERENCES: [453] – [461]

6.2   Avalon

Main Goal

Avalon is a high-level language support for distributed systems. AVALON/C++ is a set of language extensions to the object-oriented language C++. It gives programmers explicit control over transaction-based processing of atomic objects for fault-tolerant applications.

Avalon provides location, access and failure transparency.

Advantages

Avalon provides features for programmers to start and end (possibly nested) transactions, create and call servers (through RPC), and construct user-defined atomic data types that support a high degrees of concurrency.

Description

Avalon runs on the Camelot transaction facility (Sect. 8.11) which runs on top of the MACH distributed operating system (Sect. 4.13) using its RPC facility. Currently it runs on VAXs, MICROVAXs, SUNs, IBM RTs, and PMAXs as a new layer.

A *strongbox* security library is available at runtime, orthogonal to the language itself. Avalon uses type-specific locking and provides transaction semantics as well as a stable storage feature.

Replication is user-definable.

Miscellaneous

Avalon/C++ is given away by request. However, to run it, you need MACH and Camelot (and appropriate licenses) and to obtain the source, you need an AT&T source license for C++.

CONTACT: For technical issues contact Jeannette Wing

e-mail: wing@cs.cmu.edu

and for administrative and maintenance issues (getting a copy, licenses, reporting bugs,

etc.) contact Karen Kietzke, School of Computer Science, Carnegie Mellon Univ., Pittsburgh PA 15213-3890, USA
e-mail: ky@cs.cmu.edu
REFERENCES: [462] – [464]

6.3 DAPHNE

Main Goal

DAPHNE stands for Document Distributed Applications Programming in a Heterogeneous Network Environment. Its main goal is to support the distributed execution of modular programs (currently Modula-2 programs) in heterogeneous networks, i.e., networks with different hardware as well as different operating systems. To run DAPHNE, there is no need for modifications of the system's software, such as operating system, compiler, or binder. A UDP-based portable runtime environment is provided to allow calls to remote modules in a location and access transparent way. Furthermore, it supports program development through tools, such as a stub-generator or a configurator.

Advantages

The main advantage of the DAPHNE approach is the development of distributed application systems for heterogeneous environments without using special "distributed" programming languages and without modifications of the existing system software.

At present, only sequential systems are supported. There is no support of threading.

Description

DAPHNE is a heterogeneous network operating system with a Modula-2 interface. Runtime support is given in the user address space.
It runs on SUN 3 and SUN SPARC under UNIX. Ports to VAXstations under VMS and IBM/PCs under MS-DOS are under development.

The UDP protocol suite is used as a basis for a new portable remote procedure call protocol; at-most-once semantics are guaranteed.

At the time of the configuration of a distributed program, for all "remote modules" it is determined whether they are localized by a name server, localized through interaction, or looked up in a special file system. If they are looked up in a special file system, system and file names will be provided. In either case, the modules will be dynamically loaded, if and when they are needed.

DAPHNE uses the security mechanisms of the underlying operating systems. If these security mechanisms are missing, for example in a PC-environment, DAPHNE ascertains that a protected system cannot be passed without prior authentication. In this case, a password is demanded.

DAPHNE detects and deletes orphans.

Miscellaneous

OTHER ISSUES: Currently a successor project called HERON is under development. In comparison with DAPHNE, HERON will support concurrent, object-oriented systems. As a reference language, a concurrent extension of Eiffel will be used.

STATUS: At present, research is focused on a distributed development environment that comfortably integrates DAPHNE tools and itself is a non-trivial DAPHNE application.

CONTACT: Klaus-Peter Löhr or Lutz Nentwig, Freie Univ. Berlin, Fachbereich Mathematik, Inst. für Informatik, Nestorstr. 8–9, D-1000 Berlin 31, Germany

e-mail: lohr@fubinf.uucp or nentwig@fubinf.uucp

or Joachim Müller, Univ. Bremen, Studiengang Informatik, Postfach 33 04 40, D-2800 Bremen 33, Germany

e-mail: jmu@informatik.uni-bremen.de

REFERENCES: [465], [466]

6.4 DASH

Main Goal

DASH is a project at the University of California, Berkeley, and the International Computer Science Institute, Berkeley. The main goal of the project is to design system level abstractions to incorporate digital audio and video in today's distributed workstation environment. While the project started with designing its own experimental kernel, the focus of the work has now shifted to incorporate the DASH abstractions into existing systems.

Advantages

DASH defines a workload and scheduling model, the *DASH Resource Model*, as a basis for reserving and scheduling resources needed in the end-to-end handling of streams of digital audio and video. Reservation and scheduling does not just include CPU's, but other resources such as disks and networks, enabling the system to give complete performance guarantees for the handling of this data.

Due to the current status of research and its experimental character, there is a lack of integration of the resource model with a mature operating system.

Description

In the current state of the project, DASH does not constitute an operating system kernel by itself, but a set of abstractions to be added to a host operating system. Call semantics, naming conventions, security mechanism etc., therefore, do not depend on DASH but on the host operating system. The current host operating system is Mach. The DASH abstractions are implemented in the object-oriented language C++. The target machines include SUN 3/50's and SUN SPARCstations.

In conjunction with the DASH Resource Model, DASH defines a *Session Reservation Protocol* (SRP) as a basis for resource reservation over the network. SRP allows to obtain performance guarantees for IP-based communication. It can be seen as an internetwork layer management protocol that does require any changes to the TCP/IP protocol definitions, but changes the behavior of host and gateways when forwarding IP packets.

Miscellaneous

STATUS: active.
CONTACT: David P. Anderson, Computer Science Division, Dept. of Electrical Engineering and Computer Science, Univ. of California, Berkeley CA 94720, USA
anderson@charming.berkeley.edu
REFERENCES: [467] – [474]

6.5   Emerald

Main Goal

Emerald is an object-oriented language for distributed programming, as well as a run-time system supporting that language. Emerald seeks to simplify distributed programming for the application writer while also providing efficient execution. The key features of Emerald are (1) a single, semantic model for objects of all sizes, both local and distributed, and (2) on-the-fly object mobility, supported at the language level. Emerald provides access and location transparency, while supporting high-performance local object invocation through compiler and language support.

Advantages

An advantage of Emerald is its ability to create small (e.g., integers) and large (e.g., compilers) local and distributed objects using one unified object definition mechanism. When compiling an object, the Emerald compiler tries to determine how that object will be used. Different implementations may be chosen for an object based on the generality it requires, as determined at compile time. Distribution is invisible from the point of view of inter-object communication; i.e., all invocations are location-independent and the run-time system is responsible for locating the target of an invocation. However, programmers can explicitly use distribution through the object mobility mechanism that allows objects to be placed or moved for reasons of performance or reliability.

Description

A prototype implementation for a compiler of the Emerald language and a small run-time system have been constructed on networks of DEC VAXs and SUN 3s using a 10 Mbit/second Ethernet local area network.

Objects are the units of programming and distribution, and the entities between which the communication takes place. Each object exports a set of operations and can be manipulated only by the use of these operations. All information is encapsulated in specific objects.

Objects communicate through invocation. Since parameters are objects themselves, the method of parameter passing is *call-by-object-reference* (similar to CLU, Argus, Sect. 5.3), even for distributed objects. In Emerald, objects are mobile (process migration in non-object systems). This mobility enables Emerald to move parameters to the callee (*call-by-move*) so as to avoid remote references (Argus needs parameter passing by value) when desirable for performance reasons.

Emerald is strongly typed. Every variable must have an *abstract type* that defines its interface, i.e., the operation names and the number and types of parameters for invocations of objects assigned to that variable. Only valid assignments can be made: the compiler and run-time system verify that the abstract type of any object assigned to a variable *conforms* to the declared abstract type of that variable.

Emerald supports various types of polymorphism. While Emerald supports inheritance in the abstract type domain, it does not support inheritance of implementation, but work on a new model for implementation sharing within Emerald has been done.

Emerald supports inherent parallelism and concurrency, both among objects and within an object. Processes have exclusive access to objects through monitored variables and may synchronize using traditional condition variables.

Availability issues: when an object moves, it leaves behind it a *forwarding address* so that invocations sent to the old location can be forwarded correctly. If a machine crashes, forwarding addresses are lost. Emerald solves this problem by an exhaustive search for this object, implemented as a *reliable broadcast protocol*.

Security issues, replication and failure handling are not mentioned due to the current status of the project. Currently, only immutable objects may be replicated.

Miscellaneous

STATUS: A prototype implementation has been constructed. An Emerald compiler has also been written in the Emerald language. Two follow-ons have been built: Presto and Amber (Sects. 8.50, 8.3 resp.). Presto is an object-oriented system for parallel programming on shared-memory multiprocessors. Amber provides for distributed parallel programs using networks of multiprocessors. It has lightweight threads and Emerald-style object mobility. Both Presto and Amber are written in C++.

CONTACTS: Hank Levy, Dept. of Computer Science and Engineering, FR-35, Univ. of Washington, Seattle WA 98195, USA
e-mail: levy@cs.washington.edu
Norm Hutchinson, Dept. of Computer Science, Univ. of Arizona, Tucson AZ 85721, USA, or Eric Jul, DIKU, Univ. of Copenhagen, Universitetsparken 1, DK-2100 Copenhagen, Denmark.
REFERENCES: [475] – [480]

6.6   Enchère

Main Goal

Enchère is a distributed electronic marketing system consisting of loosely-coupled work-stations that communicate via messages.

Enchère provides access, concurrency, replication, and failure transparency in terms of atomic transactions.

Advantages

The main advantage of Enchère is its hardware stable storage implementation.

The main disadvantage is its limitation to dedicated applications and its special-purpose character.

Description

Enchère runs on Intel 8086 and 8085 microprocessors; machines that have been especially designed for the Enchère-system.

It is a new kernel built from scratch on the bare hardware.

Enchère uses a packet switching communication protocol. It does not provide any RPC facility.

Transparent naming is not provided, therefore Enchère does not provide location transparency.

Enchère implements hardware capabilities in a so-called Stable Storage Board. These Stable Storage Boards were also developed to guarantee the reliability of transactions. Transaction can be nested.

Process migration or any other form of mobility is not supported.

Miscellaneous

STATUS: The orientation of current research is to extend Enchère to a general-purpose, object-oriented system, the Gothic system (Sect. 5.10).

CONTACT: Michel Banatre, INRIA, IRISA, Campus de Beaulieu, F-35042 - Rennes Cedex, France

e-mail: banatre@irisa.fr

REFERENCES: [481]

6.7   Galaxy

Main Goal

Galaxy aims to be a fully transparent distributed operating system in a large scale network.

Advantages

In addition to all levels of transparency for users with message-based access to the objects, programming-level transparency on distributed virtual memory is also planned to be provided.

However, a great gap exists between the goal and the present state.

Description

A first and very simple Galaxy-prototype runs on two IBM RT PCs connected by an Ethernet. It is a new kernel where some of the low-level codings are based on that of Acis 4.2. Acis 4.2 (Academic Information System) is an IBM version of 4.2 BSD. Galaxy's communication protocol is new and built from scratch. No semantic guarantee is provided so far.

Galaxy uses an Uniform Address Space approach, i.e. the collection of memory of all nodes is treated as a single paged virtual memory space. On demand pages may migrate from one node to another. Access to remotely located pages causes a network page fault. Its page fault mechanism is built on top of the object location scheme. Each object in Galaxy has an Unique IDentifier (UID). A user-defined name of an object is translated to the UID by a system-wide naming directory which is managed distributedly. A UID is translated to the physical information such as location of the object by an IDTable. Both active and passive objects have encrypted lists of "keys." Access is allowed when the keys coincide.

In the future files will be replicated, but policies on transactions and concurrency control are not yet agreed. Processes will migrate transparently. Adaptive load sharing will be performed.

Miscellaneous

OTHER ISSUES: Many concepts and mechanisms such as hierarchical object groups (grouping objects as a new object with group-operations), dynamic replicas (partial and temporal replication of files for performance improvement), variable weight processes (continuity between normal and lightweight processes) and microprocesses (coroutines with high freedom) are proposed.

STATUS: Galaxy is a project at its very early stage. Even the main design is not completed and little of the idea is published or implemented yet.

CONTACT: Mamoru Maekawa or Hyo Ashihara, Dept. of Information Science, Faculty of Science, University of Tokyo, Hongo 7-3-1, Bunkyo-Ku, Tokyo, Japan
e-mail: ash@is.s.u-tokyo.ac.jp

REFERENCES: [482] – [484]

6.8   Global, Active and Flexible File Environment Study (GAFFES)

Main Goal

The Global, Active and Flexible File Environment Study (GAFFES) is a study of a globally distributed file system designed to share information in a world-wide network consisting of more than a million workstations. GAFFES, designed at the University of California in Berkeley, will provide location, access, concurrency and replication transparency to a huge user community.

Advantages

GAFFES is highly distributed in order to promote high capacity and modular growth. It provides independent services such as naming, replication, caching, security, and authentication. GAFFES is a serious approach to build a globally distributed file system in a huge world-wide communication network.

GAFFES is an experimental design. It neither has been built, nor is it intended to be built.

Description

GAFFES could run on future networks varying in speed and architecture, such as low-speed telephone lines and high-speed optical fibers. The client workstations connected by this network would be quite heterogeneous, ranging in power from PCs to large mainframes and running many different operating systems. The heterogeneity extends into administration. The network crosses national boundaries and is managed by different organizations resulting in difficult security and privacy issues.

GAFFES was designed to be built on a base of traditional file servers and RPC communication. RPC communication can be performed between any two machines in the network.

GAFFES files have unique names which are independent of a user's location or the file's location or the number of replicas. Moreover, clients may identify files with descriptive names allowing file accesses based on their contents. The GAFFES name space is hierarchical. Each component of a file name further qualifies the object in the name space.

The file system is designed to provide a version history of files. Any kind of manipulation is supported on the version tree. New versions are created with each modification, and the naming mechanism ensures that clients always access the latest version as far as *unqualified* requests are made. Unqualified requests do not contain a version number for any particular file.

Both client and server machine have a globally unique identifier. Both contain hardware encryption devices which can be used in security procedures. On demand, public-key/private-key pairs can be used for secure data transmission. Each organization has autonomous control over the security of its own files. The rights to user access are freely granted or revoked. A GAFFES security system provides a uniform mechanism for authenticating users and ensures the privacy of data.

The creator of a file is responsible for the degree of replication. Thereby he/she determines the availability of the new file. Each client performing a operation specifies whether caching should be used and what degree of consistency is required for the cached copies.

All client requests are performed as part of a transaction which is run as a set of atomic serializable actions. Nested transactions are supported by a traditional file server interface.

Miscellaneous

STATUS: GAFFES is just a design, not an implementation. GAFFES is a first step towards the design of a large-scale globally distributed system.
CONTACT: Douglas Terry, Xerox Palo Alto Research Center, 3333 Coyote Hill Road, Palo Alto CA 94304, USA
e-mail: terry.pa@xerox.com
REFERENCES: [485], [486]

6.9   Grapevine

Main Goal

Grapevine was developed at Xerox Palo Alto Research Center as a distributed, replicated system that provided message delivery, naming, authentication, resource location and access control services in an internet of computers. Access to the servers as well as location and replication of entries were transparent to the users.

Advantages

Grapevine was a reliable and highly available mailing system through replication of functions among several servers. However, this led in some cases to unexpected delays and errors. Several details in the implementation were changed in order to efficiently serve the proclaimed 10 000 users.

Description

The Grapevine servers ran on Alto computers and 8000 NS processors with the Pilot operating system. Client programs ran on various workstations under several operating systems throughout the Xerox network, with nodes located in clusters all over the globe.

Therefore, the communication was built on internet protocols. Either individual packets could be sent (unreliable) or byte streams could be established with the PUP protocol, which were sent reliably and acknowledged.

Each entry in the registration database had a two-level hierarchical name, the registry being the first part (corresponding to locations, organizations or applications), thus dividing the administrative responsibility, and the name within that registry as the second part. The choice of the servers was done by programs on the client side, by choosing the nearest available server.

Authentication at associated file servers was done by the use of Grapevine. In this manner the protection scheme contained individuals and groups. To improve performance, the file servers cached access control checks.

Replicated submission paths were available, because each server could deliver messages to any user and each user had inboxes on at least two servers. Registries were replicated at various registration servers as a whole. Each registry was administered independently of the others. Clients had to cope with transient inconsistencies. Any conflicting updates of registration entries did not get solved.

Miscellaneous

OTHER ISSUES: Grapevine was also used to control the integrated circuit manufacturing facilities.

STATUS: Grapevine was in use by Xerox, where the system should have consisted of up to 30 servers and 10 000 users (today, 17 servers and 1500 users in 50 local area networks). The registration service became a product of Clearinghouse service (Sect. 8.15).

CONTACT: Michael D. Schroeder, DEC Systems Research Center, 130 Lytton Ave., Palo Alto CA 94301, USA

e-mail: mds@gatekeeper.pa.dec.com

REFERENCES: [487], [488]

6.10   Heterogeneous Computer System (HCS)

Main Goal

The Heterogeneous Computer Systems project (HCS) was designed to integrate loosely-connected collections of hardware and software systems. The costs of adding new types of systems should be decreased, and the number of common facilities and high-level services that users can hope to share should be increased. The shared facilities include remote procedure call and naming, while the services include mail, filing and remote execution.

Advantages

The main advantage is its capability to incorporate a new system at low cost and without masking the unique properties for which it was obtained. Prototypes, but not products, of each facility and service have been built.

Description

Prototypes of the HCS facilities and services run on various hardware and software systems including SUN, VAX, Tektronix, and XEROX D-machines. Operating systems supported include (various versions of) UNIX and XEROX XDE.

The communication mechanism is heterogeneous remote procedure call (HRPC), which emulates the native RPCs of the underlying systems by specifying interfaces to the five principal components of an RPC: the stubs, the binding, transport and control protocol as well as the data representation. These components are separated from each other and can be selected dynamically to achieve flexibility at the cost of additional processing at bind time. The performance of the HRPC is essentially identical with that of native RPCs (tested for the SUN and Courier RPC). Language support for C, Mesa, and FranzLisp has been explored.

The name service (HNS) provides a global name space with uniform access by the use of the local name services (including BIND and XEROX Clearinghouse, see Sect. 8.15). Each name contains a context that specifies the name service and an individual name that determines the local name on that service. The name-service-specific code is encapsulated in a set of HRPC-accessible routines, which can be added easily without any recompilation of existing codes.

Neither security issues nor measurements for reliability and performance improvements play any role.

Miscellaneous

The prototype of a filing service (HFS) is structured like the name service, with file-service-specific code mediating accesses to shared files. Each source program is responsible for providing abstract type information about a file when it is accessed. A mail service has been developed that coexists with existing mail services by a *global alias* service that contains entries for each user. The HCS environment for remote execution (THERE) is a facility for constructing remote execution clients and servers in a network. Both client and server make use of service specific interpreters.

STATUS: The project is no longer active.

CONTACT: David Notkin, Department of Computer Science and Engineering, FR-35, University of Washington, Seattle WA 98195, USA

e-mail: notkin@cs.washington.edu

REFERENCES: [489] – [495]

6.11 Incremental Architecture for Distributed Systems (INCAS)

Main Goal

The INCAS project (INCremental Architecture for distributed Systems) puts its emphasis on the development of a comprehensive methodology for the design of parallel and locally distributed systems. It supports large grain parallelism at the process level.

The main goal of the project is the development and evaluation of models and tools for distributed programming as well as investigation of distributed control algorithms.

Advantages

The main advantage of the project is that it covers a wide scope ranging from hardware development over distributed operating systems up to high-level distributed applications. Thus, it addresses the entire spectrum of multicomputer-related software aspects such as distributed programming languages and environments, and distributed test methodologies.

Description

A network of UNIX workstations connected by Ethernet and a Transputer cluster serve as an experimental testbed for INCAS.

The project aims at the development of an integrated environment for general purpose distributed programming that supports the activities of design, implementation, execution, as well as behavior and performance analysis of distributed applications. The current work is based on experience gained in the earlier stages of the project where corresponding tools were built and assessed.

The central part of the INCAS project was the development of LADY (Sect. 8.37), a language for the implementation of distributed operating system, and CSSA (Sect. 8.17), a distributed application language.

The Computing System for Societies of Agents (CSSA) is an experimental high-level language for concurrent and distributed programming. It is based on the notion of *actors*. Actors are active objects that communicate with each other solely by asynchronous message passing. Its object-oriented message passing philosophy, called "module-message paradigm," is especially well suited to distributed computing.

The object-based Language for Distributed Systems (LADY) is developed for the implementation of distributed operating system. It provides broadcast and multicast communication modes at the language level. Furthermore, it allows for the modification of communication paths, and the creation and deletion of *teams* at runtime. A team is an autonomous entity and implements a set of related operating system functions. Data exchange among teams (inter-team communication) is exclusively performed by message passing using so-called export and import interfaces. Within the teams, processes and monitors are the most important building blocks. Communication among processes (intra-team communication) is accomplished via monitors.

The structuring concepts of LADY are expressed by three language levels: (a) the system level, (b) the team level, and (c) the module level. Similar concepts can be found in Argus, CHORUS, and Eden's EPL (Sects. 5.3, 5.5, and 5.9 resp.).

Miscellaneous

STATUS: A first prototype of the complete system including compilers and run-time
environments for LADY and CSSA as well as a distributed test and monitoring facility
for LADY programs has been available since 1985. The programming languages were extensively used by several research projects at the University of Kaiserslautern, Germany.
The project now aims at the development of an integrated environment for general purpose distributed programming that supports the activities of design, implementation,
execution, and analysis of distributed applications. Active.
CONTACT: Jürgen Nehmer, Fachbereich Informatik, Univ. Kaiserslautern, Postfach
3049, D-6750 Kaiserlautern, Germany
e-mail: nehmer@informatik.uni-kl.de
REFERENCES: [496] – [498]

6.12   ISIS

Main Goal

ISIS is a distributed software layer providing a procedural interface and in this way
resembles a distributed operating system. However, ISIS is built over UNIX. The system
is the base of a number of higher level application efforts included a distributed file
system, called *Deceit* (Sect. 8.21), a powerful system for developing reactive control
software, called *Meta* (Sect. 8.41), a "replicated" NFS, called *RNFS* (Sect. 8.53), a
network resource manager (for load sharing over a network of workstations) and other
similar facilities. ISIS users access the system at either a procedural level (building
their own customized servers) or at a utility level (i.e., using the *Deceit* file system).
Increasingly, ISIS users treat the basic process group construct as a form of "distributed
object" support. One implication is that the members of a group may be heterogeneous
and that one process may be in many groups.

The ISIS Project is concerned with providing tools for fault-tolerant distributed computing in the form of a uniform program-callable interface under UNIX and similar operating systems. ISIS provides a collection of distributed computing functions under
a general architecture based on support for *process groups* and *group communication*
(through group broadcasts, one-to-one remote procedure calls, and asynchronous message passing). The system is notable in providing mechanisms for building fault-tolerant
software as well as distributed systems that can reconfigure dynamically without losing
data consistency.

ISIS provides all levels of transparency, but they are optional. One can "see" and
exploit distribution if desired, or conceal it completely by treating a process group as a
single, reliable entity. ISIS is probably unique in providing concurrency transparency.
That is, ISIS allows asynchronous, highly concurrent systems to look synchronous (as if
one thing happens at a time). This is called *virtual synchrony*.

Advantages

At the level of the Toolkit itself, ISIS made several basic contributions. First, ISIS has demonstrated that an environment providing the appropriate combination of synchronization and process group mechanisms permits construction of a huge variety of distributed algorithms. ISIS has also demonstrated that fault-tolerance is really no different from any other synchronization problem.

ISIS has also demonstrated that there is a tradeoff between asynchrony in the application (needed for performance) and the form of order provided by the communication or broadcast subsystem. ISIS was one of the first projects to suggest that communication systems should enforce *causal* message orderings, a type of ordering identified by Lamport but not previously treated as something a system could implement. The ISIS *cbcast*-primitive is a powerful asynchronous multicast primitive, capable of emulating any other such primitive at minimal cost.

At the engineering level, ISIS has explored alternatives for implementing cbcast in settings with large numbers of process groups, and for scaling group-based systems while maintaining strong consistency properties. New "fast causal multicast" protocols impose minimal overhead over the message transport costs and are ideal for kernel implementations or use over hardware multicast.

The higher-level ISIS applications have also made basic contributions. *Deceit* has demonstrated that a file system can use replication for fault-tolerance without requiring complex kernel protocols or consistency preserving mechanisms to deal with partitioning. *Meta* represents a major advance in distributed application management technologies, and deals simultaneously with temporal and logical consistency issues to ensure that Lomita rules fire exactly when they should, despite clock synchronization limits, failures and asynchrony. Similar statements can be made about other high-level software.

Description

ISIS runs on a variety of systems: UNIX on SUN 3, 386i, and SPARC CPUs, IBM (PC and RS/6000), DEC (MicroVAX and MIPS 2100, 3100 and 5000), HP (68000 and Spectrum 800 series), Apple (Mac. II under AUX), Apollo, Gould. DEC VMS port is underway. ISIS is callable from C, C++, Fortran 77, Common LISP, Prolog. UNICOS (Cray) and VM/CMS (IBM) ports are expected in 1990.

ISIS is a new procedure-call layer (with additional utility programs) currently exploring options to integrate into CHORUS kernel (Sect. 5.5), MACH 3.0 network (Sect. 4.13), message server, and Amoeba system (Sect. 5.2).

ISIS uses UDP and provides point-to-point RPC and IPC as well as group multicast and broadcast with asynchronous, 1-reply, k-reply, all-reply, and majority-reply options.

Virtual synchrony is a powerful order-based property enabling asynchronous systems to run "as if they were synchronous", with failure atomicity and all necessary synchronization done by ISIS, transparently. Most communication is at-most-once, atomic with regard to crash failures and message loss, and atomically addressed (i.e., a message goes to all current members of a process group and this is rigorously defined).

ISIS uses a hierarchical name space that looks like a UNIX file system, but with an optional scope specifying the LAN or set of machines on which to search for the object. ISIS is intended to hide behind such mechanisms as MACH ports and the UNIX name space (symbolically named ports) as much as possible.

ISIS has an optional security subsystem based on the Needham et al. protocols and using RSA and DES as the underlying cryptosystem. Use of the system is optional and authentication is costly (although the cost is only paid once). The basic scheme yields authenticated knowledge of the user-Id of a process with which one is communicating (or of which a group is composed) and permits the creation of authenticated channels for extended communication sessions. The high cost of ISIS is the cost of authenticating a remote-Id or opening such a channel; on a DEC 3100 workstation this takes about 5 seconds. When not using this, ISIS offers no special security beyond that common in UNIX. But, it never shows the contents of a message to any process except the destination of that message.

ISIS provides a range of synchronization and concurrency control mechanisms including a fast replication scheme, based on asynchronous updates and token passing. It supports transactions as an option but is normally based on virtual synchrony, the communications-oriented form of serializability in which individual communications events and group membership changes are serializable but there are no higher-level transactions. This approach affords high performance, much better than for transactions on replicated data. Moreover, it lets ISIS support a wide variety of algorithms for cooperation on a group basis: subdividing computational tasks, monitoring, etc.

ISIS handles crash failures (servers and clients) and will not allow inconsistencies to form if the network partitions (however, it may shut down one side of the partition). ISIS supports optional stable storage mechanisms. Orphan detection is not relevant since ISIS is not transactional.

ISIS does have a powerful group join mechanism that allows transfer of application-specific state information. "Migration" (not really a process migration scheme) is done by state transfer followed by termination of the "old" process. But this is not a completely transparent mechanism. In particular, the application programmer must define the "state" to be transferred.

Miscellaneous

OTHER ISSUES: As mentioned, ISIS is not a distributed operating system, but rather a software layer that extends an underlying operating system to support new functionality. This underlying system might be a BSD or System-V variant of UNIX, MACH, or even VMS. On the other hand, ISIS does provide a functional interface suitable for developers of new distributed or fault-tolerant system services. For example, ISIS has been used to develop transparently fault-tolerant file systems that support file replication, while presenting the user with a normal NFS interface. The file system called *Deceit* is just one of many ISIS applications. Others include the *Meta* system (for instrumenting a distributed system and developing reactive control software) or the RNFS approach (Sect. 8.53). Furthermore, a parallel make program, a network resource manager, a network "data publication" service, and a transactional subsystem for database integration have been supported. ISIS thus spans a range of functionality from the procedure-call level at which process groups and group-oriented tools are implemented up to high-level subsystems with sophisticated distributed functionality of their own.

The lowest layer of ISIS defines the process group interface and *toolkit*. Tools are included for creating, joining and leaving process groups, obtaining a capability on a group, multicasting to a group (either synchronously or awaiting some number of responses), IPC and RPC to individual processes, replicated data management with synchronization based on token passing (like locks), division of a computation among groups of processes (several application-specific styles), monitoring groups or individual processes, transferring state atomically to a new group member, saving state persistently, performing database transactions, spooling data for a service that runs periodically, and long-haul communication (i.e., to a remote LAN). These tools are comparable to the UNIX system calls in sections II and III of the UNIX manual. In all cases, the designer can program for a simplified "synchronous" execution model and will benefit from the maximum possible asynchrony. This transformation is transparent (concurrency transparency). The approach allows high performance and lets applications maintain consistency of group state even when failures occur. A higher level of ISIS offers fault-tolerant applications. STATUS: ISIS V2.1 is now available in source form without fee or restriction.

An enhanced commercially supported version of ISIS is being marketed by ISIS Distributed Systems, Inc, 111 South Cayuga Street, Ithaca NY 14853, USA

ISIS has been distributed to more than 500 commercial and academic groups worldwide and is actively used in dozens of settings. CONTACT: Ken Birman, Dept. of Computer Science, Cornell Univ., 4105 Upson Hall, Ithaca, NY 14853, USA e-mail: birman@cs.cornell.edu or isis@cs.cornell.edu REFERENCES: [499] – [505]

6.13   Medusa

Main Goal

Medusa was a distributed operating system designed for the Cm* multimicroprocessor. It was an attempt to produce a system that is modular, robust, location transparent, and to take advantage of the parallelism presented in Cm*.

Advantages

Each Cm contained an LSI-11 processor and 128 kB of main memory. This hardware constraint lead to the following design decision. Medusa discarded the assumption that every processor in the system must be able to fulfill all services within the operating system. The entire operating system was therefore divided into disjoint utilities resulting in sometimes "unnecessary" non-local executions. On the other hand, utilities were a good example of how an operating system can be distributed as a whole.

Description

Medusa was the successor to the StarOS operating system (Sect. 8.61). StarOS was the first operating system developed at Carnegie Mellon University for the Cm* multimicroprocessor computer. Medusa's functionality was divided into disjoint *utilities* due to the small main memory of the Cm internal LSI-11 processor. Just a small interrupt handling kernel was distributed among all processors.

There was no guarantee given that any particular processor contained a copy of any particular utility. Besides, a processor was only allowed to execute a local residing utility. In Medusa, messages therefore provided a *pipe*-based mechanism for synchronous (point-to-point) cross-processor function invocations. Message pipes contained no information as to location. Each utility protected its own execution environment. It might migrate around the system, once the format for message exchange via pipes has been selected.

Medusa provided an object-oriented model. Each object was manipulated by type-specific operations. Access to objects was protected by descriptor-lists. An important object class was the semaphore class which provided concurrency control for shared objects. Deadlock detection was supported.

Concurrency support was achieved by the so-called *task force* concept. This pertained to a collection of concurrent *activities*, entities that actually got scheduled for execution (analogous to processes in traditional systems). All activities of a given task force could run simultaneously (almost all the time).

Failure handling, replication and availability issues were not addressed.

Miscellaneous

STATUS: The design for Medusa was begun in the late Fall of 1977, and coding started in the summer of 1978. Not active.
CONTACT: John Ousterhout, Computer Science Division, Univ. of California, Berkeley, 571 Evans Hall, Berkeley CA 94720, USA
e-mail: ouster@arpa.berkeley.edu
REFERENCES: [506] – [509]

6.14   Meglos

Main Goal

Meglos extends the UNIX operating system with simple and powerful communication and synchronization primitives needed in real-time environments. Location and access transparent communication is achieved using channels.

Advantages

The main advantage of the Meglos operating system is that it can be used in real-time environments. The major constraint for a Meglos user is the hardware as described below.

Description

The Meglos operating system is implemented as a new, UNIX-extending kernel. Meglos consists of up to 12 processors using M68000- based Multibus computer systems. Each processor is connected to the 80 Mbit/second S/Net. A DEC VAX computer running UNIX serves as a host processor for the system. The satellite processors are built around the Pacific Microsystems PM-68k single-board computer.

Meglos is an exact emulation of the UNIX environment allowing programs to make use of I/O redirections and pipes. Programs running under Meglos are controlled from the UNIX system which provides access to files.

A specialized IPC has been designed to support real-time applications providing at-most-once semantics. The IPC mechanism is based on communication paths called *channels*. Each channel permits bilateral communication via message exchange (*rendezvous-* concept). Meglos supports multicast communication as well. Independent flow-control is provided for every channel to meet real-time constraints. Channels are named and provide location as well as access transparent communication, as well as synchronization between readers and writers. For real-time response, a priority-based preemptive scheduler and an efficient process switching has been implemented.

Security issues, failure handling and availability of the system are not addressed.

Miscellaneous

OTHER ISSUES: MIMIC, a robot teaching system has been implemented under Meglos using real-time features. Also, Meglos includes a symbolic debugger for C programs.
STATUS: Several prototypes of Meglos systems have been built and are being used to support multiprocessor robotics applications in the AT&T Robotics Research Department. The current status is unknown.
CONTACT: Robert D. Gaglianello, AT&T Laboratories, Holmdel NJ 07733, USA
REFERENCES: [510] – [513]

6.15 Mirage

Main Goal

Mirage and Mirage+ are prototype distributed shared memory (DSM) systems. The goal of Mirage and Mirage+ is to study the feasibility of distributed shared memory in a loosely coupled environment. Mirage provides coherency with concurrent write-sharing.

Advantages

Mirage's time-based coherency provides distributed sharing with a minimum of network traffic. Other advantages include transparency, ease of use, and a simplified programming model.

Disadvantages include lack of recovery in the case of site failures, and a potential lack of scalability. The development of Mirage+ is an attempt to address the deficiencies of Mirage as well as other issues.

Description

Mirage is a DSM system implemented in the kernel of a UNIX-based operating system. Mirage has been implemented under UNIX System V on a system of 3 VAX 11/750s networked together using a 10 Mbit/second Ethernet.

Processes in a DSM network may logically access the address space of remote processes. DSM systems provide an efficient method of locating, retrieving, and insuring the consistency of data located across the network.

The access protocol used in Mirage has two distinctive features: First, the model of sharing is based on paged segmentation, and second, the user of a shared page is guaranteed a minimum time of uninterrupted access. Access to a shared page is controlled by the page's "library site". The library site keeps track of outstanding copies, including the most recent version of the page. The location of shared pages is transparent to users, whose page copies are appended to their virtual address space. Remote locations and names are found using a special purpose naming protocol.

Miscellaneous

Mirage+, currently under development, is an enhancement of the original Mirage system. Major focuses include scalability, reliability and optimum page and process placement. Page accesses in Mirage+ will be modeled as transactions in order to preserve data integrity in case of a singe site failure. Process and page migration will be used to increase load balancing and to minimize the number of page faults due to sharing.

Mirage and Mirage+ are being investigated in a homogeneous network environment. There are no current plans to support heterogeneity.

STATUS: Mirage+ is currently under development on a network of 16 PS/2 workstations using a 10 Mbit/second Ethernet.

CONTACT: Brett D. Fleisch, Computer Science Department, Tulane University, New Orleans LA 70118, USA

e-mail: bdf@cs.tulane.edu

REFERENCES: [514] – [521]

6.16   Network Computing System (NCA/NCS)

Main Goal

The Network Computing System (NCS) is a portable implementation of the Network
Computing Architecture (NCA), a framework for developing distributed applications.
It supports various machines running different operating systems. It is designed to
represent the first step in a complete network of a computing environment by providing
all aspects of transparency.

Advantages

NCA/NCS eases the integration of heterogeneous machines and operating systems into
one computing environment thus expanding the set of network applications. A comparison with similar systems, such as HCS (Sect. 6.10), is difficult since performance has
not been measured.

Description

NCS currently runs under Apollo's DOMAIN/IX, UNIX 4.x BSD and SUN's version of
UNIX, as well as under MS/DOS and VMS. The system supplies a transport-independent
remote procedure call facility using BSD sockets as the interface to any datagram facility, for example UDP. It provides at-most-once semantics over the datagram layer, with
optimizations if an operation is declared to be idempotent. Interfaces to remote procedures are specified by a Network Interface Definition Language (NIDL). NCS includes a
portable NIDL compiler which produces stub procedures that handle data representation
issues and connect program calls to the NCS RPC runtime environment. A forthcoming
version (NCS 2.0) will run over both, connection-oriented and connectionless transport
protocols.

  Objects are the units of distribution, abstraction, extension, reconfiguration and
reliability. A replicated global location database helps locate an object, given its universal
unique identifier (UUID), its type or one of its supported interfaces. UUIDs are an
extension of DOMAIN's UID (Sect. 3.1).

  A Distributed Replication Manager (DRM) supports a weakly consistent and replicated database facility. Weak consistency means that replicas may be inconsistent but,
in the absence of updates, converge to a consistent state within a finite amount of time.
A simple locking facility supports concurrent programming.

Miscellaneous

OTHER ISSUES: A Concurrent Programming Support (CPS) provides lightweight tasking facilities. A data representation protocol defines a set of data types and type con-
structors which can be used to specify sets of typed values which have been ordered.
Version NCS 2.0 will use the POSIX draft multi-thread API (Pthreads) and not the
CPS.

STATUS: Apollo has placed NCA in the public domain. Implementations are currently
in progress for further machine types. A future version should include a general purpose
name server that will be able to find an object by attributes rather than by text name.
A transaction facility and strongly consistent replication are additionally planned.

CONTACT: Nathaniel W. Mishkin, Hewlett-Packard Company, Chelmsford MA 01824,
USA

e-mail: mishkin@apollo.com

REFERENCES: [522] – [524]

6.17   Plan 9

Main Goal

Plan 9 is a distributed computing environment currently being developed at the Bell
Laboratories. It is a general-purpose, multi-user, portable distributed operating system
implemented on a variety of computers and networks.

It is an operating system with physically distributed components. Separate machines
acting as CPU servers, file servers, or terminals are assembled and connected by a single
file-oriented protocol as well as local name space operations. Its main goal is to provide
levels of efficiency, security, simplicity, and reliability seldom realized in other distributed
system approaches.

Advantages

Plan 9's simplicity and efficiency is the result of putting function where it is best performed. The file server is a file server and nothing else, i.e., not a UNIX system that also
serves as a file system. The terminals are tuned to be user interfaces and not compile
engines, though they can perform the latter function if CPU servers are unavailable. All
of the components combine to offer the user what seems to be a very powerful UNIX-
like system, usually at a much lower cost than an equivalent collection of independent
workstations.

However, it lacks a number of features often found in others distributed systems, such
as a uniform distributed name space, process migration, and lightweight processes.

Description

Currently the production CPU server is a Silicon Graphics Power Series machine with
four MIPS processors connected to the file server via DMA connection. Datakit and
Ethernet are used to connect it to terminals and non-Plan 9 systems. The file server
is another Silicon Graphic computer with two processors (to be replaced by a MIPS
6280 single-processor machine). It provides location transparency to the clients. Gnot,
a locally designed machine, runs as standard terminal for Plan 9.

Plan 9 has two kinds of name space: the global name space of the various server machines, and the local name space of files and servers visible to processes. Since Datakit provides naming for its machines, global naming need not be handled by Plan 9.

Security issues are not directly addressed. Currently a simple authentication manager on the Datakit network is used. It is planned to replace the authentication manager by a Kerberos-style system.

Although Plan 9 allows to instantiate processes where they should run, it does nothing to make process migration happen.

Miscellaneous

In Plan 9 the user relies on the distributed components rather than having a self-sufficient little box on his desk. The designers have packaged some systems for home use that have their own internal file system and can act as CPU servers when disconnected from the rest of Plan 9. However, these are no better or different than the typical workstation. They have all the administration problems that typical workstations have. Work is in progress to turn these local disks into caches of the "file servers" to remove some of these problems.

STATUS: A complete version of Plan 9 was built in 1987 and 1988. In May 1989 work was begun on a new version. This version was based on the SGI MIPS-based multiprocessors and used the first version for the bootstrap. The SGI file server became operational in February 1990. Ports to other hardware are in progress.

CONTACT: Dave Presotto, AT&T Bell Laboratories, Murray Hill NJ 07974, USA
e-mail: presotto@research.att.com

REFERENCES: [525]

6.18   Psyche

Main Goal

Psyche is an operating system for shared memory multiprocessors. It is currently under development at the University of Rochester. From the point of view of individual applications, the Presto system (Sect. 8.50) is a close relative to Psyche. However, Psyche is an operating system, Presto is a user-level runtime library. Psyche is an attempt to find the right interface for supporting many different models of process and communication, Presto is an attempt to allow the user to customize what the operating system already exports.

Advantages

Psyche is an operating system designed to support multi-model parallel programming on large-scale shared memory multiprocessors. It provides mechanisms ensuring that threads created in user space can use the full range of kernel services, without compromising the operations of the peers. Data sharing is the default, not the exception. Access rights are distributed without interaction with the kernel. Psyche presents an explicit (user definable) tradeoff between protection and performance.

Description

Psyche is a new kernel built on the bare machine. It is designed to run on the BBN Butterfly Plus multiprocessor, based on the GP1000 product line.

Psyche makes use of the Ethernet interface for network file services and communicates via UDP.

Psyche supports *multi-model parallel programming*. Multi-model programming is the simultaneous use of different models in different programs and combinations of models in the same program. Shared memory and message passing are two different communication mechanisms, each of which could be used in programs having nothing to do with each other. Alternatively, the models employing them could be connected together if the task at hand required it. For example, the Psyche people are building a checkers-playing robot that uses four different models: (1) a high-level planner, written in Lynx that does parallel alpha-beta game tree search, (2) a low-level vision processing module that recognizes human moves, (3) a low-level motion planner written in MultiLisp that figures out where to move a robot arm to reflect the computer's last move, and (4) various service modules written in uthread, a Presto-like thread package. Models are connected together using procedure calls which may or may not cross protection boundaries.

An abstract data object called *realm* is the fundamental kernel abstraction. It can be used to provide monitors, RPC, buffered message passing, and unconstrained shared memory. A realm consists of code and data. The code contains the operation (and the protocol description) by which the data are accessed. Due to the encapsulation of code and data, interprocess communication is effected by invoking operations of realms. Realms have system-wide virtual addresses (uniform address space). The Psyche kernel maintains mapping from virtual addresses to realms.

Sharing is the key concept in Psyche. However, users can explicitly ask for protection if they are willing to "pay" for it in terms of performance.

Psyche is designed for migration and replication of objects. Thereby, it should maximize locality. However, data migration and replication are backed out of the current implementation.

Miscellaneous

OTHER ISSUES: If a more sophisticated approach to process management and message passing is preferred, the communication library and link realms can easily be modified to provide the communication semantics of other distributed programming languages. A wide range of process and communication models can easily be implemented on top of Psyche, for example Lynx style links (Sect. 8.38) instead of Charlotte style links (Sect. 4.3), or a Linda style shared global buffer (Sect. 8.59) instead of pure shared memory.

STATUS: Implementation began in the summer of 1988. A first toy program ran late in 1988. The first major application was ported to Psyche in November 1989. Work on

Psyche is still in progress.

CONTACT: Michael L. Scott or Thomas J. LeBlanc or Brian Marsh, Dept. of Computer
Science, Univ. of Rochester, Rochester NY 14627, USA
e-mail: {scott,leblanc,marsh}@cs.rochester.edu
REFERENCES: [526] – [529]

6.19   Server Message Block Protocol (SMB)

Main Goal

The Server Message Block Protocol (SMB) was developed to support programmers who
design or write programs that interact with the IBM PC Network Program via the
IBM PC Network. Netbios and the Redirector, software elements of the DOS operating
system, can be combined with SMB to provide most of the features found in distributed
file systems. SMB provides access transparency.

Advantages

The main advantage of SMB is the fact that a huge user community – PC users under
DOS – are enabled to take part in the world of distributed file systems. The final
development of the features of a DFS is still in progress and the disadvantage of not
supporting crash recovery, transactions and security issues will be eventually eliminated
due to a "critical mass" effect. The commercial character of this product and the number
of installed systems involved will lead to a fast enhancement of SMB.

Description

SMB is the most visible methodology used in commercial networks, for example Microsoft's MS-Net and IBM's PC LAN Program. The native operating system of SMB
is DOS (such as UNIX for NFS, Sect. 2.8, or RFS, Sect. 2.9). SMB has thus far been
implemented on UNIX V, UNIX 4.x BSD, and VMS. Applications running on a variety
of PCs make standard system calls to DOS. These calls are intercepted by the so-called
Redirector which decides whether these calls require remote access or not. Given that a
remote access is necessary, the call is transferred to the Netbios software. Then Netbios
establishes a reliable VC connection with a remote server. This server runs a special software which includes the share mechanism. Share is a main component of DOS. Share
allows opening of remote files with various permissions. DOS semantics of the entire
client file system is maintained.

The well known mounting mechanism allows expansion of a local file system by
mounting remote file system branches on different file volumes. A hierarchical name
space is the result.

SMB does not support any authentication mechanisms due to the fact that SMB
was first created in a DOS environment in which security plays a minor role. However,

authentication at the level of file system is possible by means of passwords for access to remote resources.

SMB uses a stateful protocol (like RFS). Detecting crashes is a problem and recovery mechanisms have not been defined.

File locking for concurrency control is provided.

Miscellaneous

STATUS and CONTACT: Microsoft's MS-Net and IBM's PC LAN Program are being marketed by Novel (Netware), 3COM (3+).
REFERENCES: [530] – [532]

6.20   Symunix

Main Goal

The Symunix Operating System is not what one normally considers a "distributed" system, rather, it is a tightly coupled shared memory system, with highly parallel operation of Ultracomputers and other large scale shared memory MIMD multiprocessors. "Transparency" takes on a different meaning for this kind of system, in which all memory is equally shared by all processors. Processes may run on any processor, and generally don't care which one, so it may be said that Symunix has location and access transparency.

Advantages

Highly parallel operation is possible by avoiding serial bottlenecks caused by critical sections. This allows scaling to a very large number of processors (hundreds or thousands).

However, Symunix does not address traditional distributed system problems of networking.

Description

Symunix, a heavily modified UNIX kernel, makes use of a "self service" approach, where processes perform services on their own behalf as much as possible. Where this is not possible in user mode, the process may enter the kernel (i.e., make a system call), and perform the operation there. Critical-section-free synchronization algorithms are employed so that independent processes can operate concurrently on shared data.

Symunix does not address heterogeneity. The target is large shared memory MIMD multiprocessors, such as the NYU Ultracomputer and the IBM RP3.

Like Symunix-1, Symunix-2 is fully symmetric. There is no processor which plays a distinguished role.

Symunix likes to put names in the filesystem name space. For example, all process memory consists of mapped files.

Semantics and security issues are full UNIX, plus additional features.

The scheduler will normally migrate processes at will from processor to processor; this is completely transparent.

Miscellaneous

STATUS: Symunix-1, based on UNIX V7, is running in production on an 8-processor-prototype since 1984. Symunix-2, a much more ambitious system, is still in development. Symunix-2 will support UNIX 4.3 BSD protocols.
CONTACT: Jan Edler, NYU Ultracomputer Research Laboratory, New York Univ., 715 Broadway, 10th floor, New York NY 10003, USA
e-mail: edler@nyu.edu
REFERENCES: [533] – [537]

6.21   Synthesis

Main Goal

Synthesis is a distributed and parallel operating system, integrating uniprocessors and multiprocessors over LANs and WANs. Synthesis has its own native distributed file systems, but will also emulate other popular DFSs.

The Synthesis philosophy is that every resource should be accessible in useful ways, whether transparent or not. So Synthesis provides different degrees of transparency. The object-Id does not imply a physical address, but its location can be found if required. So the Synthesis kernel provides location and access transparency. Concurrency, replication, and failure transparency depend on the particular facility. Concurrency transparency is provided for particular kernel objects, for example, multi-producer and multi-consumer queues. But Synthesis also provides queues for 1 producer and 1 consumer, which do not support concurrency. In general, Synthesis provides useful services, without imposing particular solutions.

Advantages

The key idea in Synthesis performance is the use of a code synthesizer in the kernel to generate specialized (thus short and fast) kernel routines for specific situations. There are three methods to synthesize code: first, Factoring Invariants to bypass redundant computations and second, Collapsing Layers to eliminate unnecessary procedure calls and context switches, and finally Executable Data Structures to shorten data structure traversal time. Applying these methods, the kernel call synthesized to read /dev/mem takes about 15 microseconds on a 68020 machine.

Synthesis kernel supports three kinds of objects: threads, memory, and I/O. These can be composed to form larger units of computation. Memory is a unit of data storage and protection. Thread is a unit of computation. I/O is a unit of data movement among hardware devices and memory units. Given these building blocks, it is planned to write emulators of other DOSs, which will run efficiently using kernel code synthesis.

Description

Currently Synthesis runs on three kinds of machines: the Quamachine, which is 68020-based homebrew; the Sony 1830 NEWS workstation, which is a dual 68030 workstation; and a Motorola VME-based 68020 machine (MVME135). The two 68020 versions are similar, but the 68030 is a significant upgrade, including virtual memory.

Synthesis is a completely new kernel written from scratch.

Synthesis is similar to x-kernel (Sect. 6.22) in this aspect. The Synthesis people tested the first implementation of TCP/IP, but they do not see any particular protocol as "the best". They are developing some of their own, but the idea is to support interoperability through emulation.

At present, Synthesis does not have an own RPC support. The initial RPC facilities will come from emulation. MACH (Sect. 4.13) and some flavor of UNIX will be the first targets for emulation. Synthesis' own communications facilities will most likely allow all different semantics since each can be useful in some practical situation.

The Synthesis kernel will support both a machine-oriented OID and human-oriented directory services. The Synthesis people anticipate many emulated operating systems running on Synthesis and therefore many different distributed file systems and naming schemes.

Synthesis is very concerned with security, but the approach is not yet formalized. The basic idea is that Synthesis uses "synthesized code" to provide services. The synthesized code is more flexible than capabilities, since it is an "object" in a sense. So instead of exporting protected "rights bits" to be interpreted by other kernels, Synthesis exports protected "methods" to be run by other kernels. Encryption will be necessary for secure transmission. Synthesis has designed special hardware components to protect the next generation of Quamachines from being tampered.

Transactions are considered to be on top of the Synthesis kernel. However, it is planned to use code synthesis for transaction support to minimize the layering overhead. A lot of work has been done for concurrency control, replication, and transactions (so far outside of Synthesis, see the references).

The Synthesis kernel is entirely resident on ROM. Kernel reboot on the Quamachine takes a fraction of a second. The stable storage and transaction support form the basis of client-level failure handling. So this issue has been postponed until the design of transaction support is done. At the system level, the stream model of I/O applied to disks allow direct access to disk (writing of commit records), eliminating the UNIX kernel buffering problem.

Synthesis threads support "checkpoint" and "restore" operations. With synthesized code, threads are easier to move (between machines of similar architecture). However, given the cost of process migration and problems with forwarding messages, Synthesis does not anticipate great performance gains in doing process migration.

Miscellaneous

STATUS: active.

CONTACT: Calton Pu, Dept. of Computer Science, Columbia Univ., New York NY 10027, USA
e-mail: calton@cs.columbia.edu
REFERENCES: [538] – [547]

6.22   x-Kernel

Main Goal

The x-Kernel is a configurable operating system kernel in which communication protocols define the fundamental building block. The x-Kernel provides the structure and tools necessary to conveniently configure distributed operating systems with precisely the functionality required in a particular situation.

Advantages

Because the x-Kernel is designed to be easily configurable, it provides an excellent vehicle for experimenting with the decomposition of large protocols into primitive building block pieces, as a workbench for designing and evaluating new protocols, and as a platform for the implementation of distributed applications and programming languages.

Description

The x-Kernel is a new kernel designed to efficiently support communication protocols. The original development used SUN 3 workstations; ports to other platforms are in progress.

The x-Kernel consists of "traditional" memory management and lightweight process management facilities as well as an architecture and support facilities for communication protocols. The exact set of protocols configured into an instance of the x-Kernel determines the characteristics of the x-Kernel. In other words, the x-Kernel is not a particular distributed operating system, but rather a *toolkit* for constructing such operating systems.

Included in the toolkit are a implementations of a number of protocols: the Arpanet (TCP/IP) suite of protocols, specialized protocols (reliable record stream, group communication), remote procedure call protocols with varying semantics (SUN RPC, Sprite RPC, x-Kernels's own), and fault-tolerant protocols.

Miscellaneous

STATUS: The x-Kernel implementation on SUN 3 workstations has gone through several iterations for the last several years. The latest version is currently being ported to the MIPS and SPARC RISC processors.
CONTACT: Larry L. Peterson, Department of Computer Science, University of Arizona, Tucson AZ 85721, USA

e-mail: peterson@cs.arizona.edu
or Norman C. Hutchinson, Department of Computer Science, University of British
Columbia, Vancouver V6T 1W5, B.C., Canada
e-mail: hutchinson@cs.ubc.ca
REFERENCES: [548] – [555]

Chapter 7

Table of Comparison

The table of comparison is given to summarize and compare the systems discussed. It should be viewed carefully, since in certain ways any categorized comparison can be misleading. However, this way an easy-to-read overview may be obtained. The table provides quick access to a large amount of highly condensed information. The entries are organized according to the criteria used to describe the systems. Sometimes, a similar issue or a comparable feature for an entry have been implemented. I mark this with a special symbol '+'. Entries put in brackets indicate an issue not yet implemented but planned, or indicate an optional feature.

| Name | Transparency | | | | |
	location	access	replication	concurrency	failure
Alpine		*			
Andrew	*	+	+	+	
Cedar	*	*	*	*	*
Coda	*	*	+	*	*
DOMAIN	*	*			
EFS	*	*			
HARKYS	*	*			
Helix	*	*			*
IBIS		*	*	*	
NFS	*	*			
RFS	*	*			
S-/F-UNIX		*			
Spritely NFS	*	*		*	
SWALLOW	*	*	*	*	*
VAXcluster		*		*	*
XDFS	*	*		*	

Table 7.1: Distributed File Systems — Part 1

Name	Heterogeneity	
	OS	CPUs
Alpine	(Cedar)	Dorado
Andrew	UNIX, MACH	SUNs, VAXs, DECstations, HP 3000, NeXT, IBM RTs
Cedar		Dorado, Dolphin, Dandelion
Coda		IBM RTs, DEC Pmax, Toshiba I386
DOMAIN	UNIX III, 4.2 BSD	Apollo
EFS	MASSCOMP RT UNIX	MASSCOMP MP
HARKYS	multiple UNIX systems	multi-vendor HW
Helix	XMS	68010-based
IBIS	UNIX 4.2 BSD	VAX-11/780
NFS	UNIXs, VMS, SunOS, Ultrix, MS/DOS	multi-vendor HW
RFS	UNIX V	multi-vendor HW running UNIX V
S-/F-UNIX	UNIX V	DEC PDP-11
Spritely NFS		proprietary HW
SWALLOW		
VAXcluster	VAX/VMS	VAX 7xx-11s, VAX/8600
XDFS	(Cedar)	Alto

Table 7.2: Distributed File Systems — Part 2

Name	Changes made		Communication			RPC-based	Connection		
	new kernel	new layer	standard protocols	specialized protocols	shared memory		VC	datagram	pipes/ streams
Alpine	*					*			
Andrew		*	NCS RPC, OSI TP4			*	*	*	
Cedar		*	FTP	NFS, XNS			*		
Coda	*			Camelot		*			
DOMAIN	*			*	*				
EFS	*			RDP		*			
HARKYS		*	TCP/IP			*			
Helix	*		OSI TLI				*		
IBIS		*	TCP/IP			*	*		
NFS	*		UDP/IP			*		*	
RFS	*		OSI TLI						*
S-/F-UNIX	*			KP			*		
Spritely NFS	*		XDR, UDP/IP			*		*	
SWALLOW		*		SMP				*	
VAXcluster	*			MSCP		*	*	*	
XDFS		*		PUP				*	

Table 7.3: Distributed File Systems — Part 3

Name	Semantics			Naming		Security			
	may be	at most once	exactly once	object-oriented	hier-archical	en-cryption	special HW	capa-bilities	mutual auth.
Alpine		*			*	*		*	*
Andrew	*				*	*			*
Cedar		*			*				
Coda	*				*	*			*
DOMAIN				*					
EFS	*				*				
HARKYS	*				*				
Helix				*		*		*	
IBIS	*				*				*
NFS	*				*	*			
RFS	*				*				
S-/F-UNIX	*				*				
Spritely NFS	*				*	*			
SWALLOW		*		*				*	
VAXcluster	*	*			*				
XDFS		*			*				

Table 7.4: Distributed File Systems — Part 4

Name	Availability				Failures				Object/Process Mobility
	synchro-nization	TA	nested TA	repli-cation	recovery client crash	recovery server crash	stable storage	orphan detection	
Alpine	+	*			+	+			
Andrew	*	+		+	+	+			
Cedar	*	*		*	*	*			
Coda		*		*	+	+			
DOMAIN									
EFS		*			*	*			
HARKYS									
Helix		*	*		*			+	
IBIS	*			*	*	*			
NFS	*				*	*			
RFS	*				*				
S-/F-UNIX									
Spritely NFS					*	*		,	
SWALLOW	*	*		*	*	*	*		
VAXcluster	*			+	*	*			
XDFS	*	*			*	*	*		

Table 7.5: Distributed File Systems — Part 5

Name	Transparency				
	location	access	replication	concurrency	failure
Accent	*	*			*
Alpha	*	*	*	*	*
Amoeba	*	*		*	*
Argus	*	*			*
BirliX	*	*	(*)	*	*
Cambridge	*	*			*
Charlotte	*	*			
CHORUS	*	*	(*)		
Clouds	*	*		*	*
Cosmos	*	*	+	+	+
Cronus	*	*	+		
DIOS	*	*			
DACNOS	*	*		*	
DEMOS/MP	*	*			*
DUNE	*	*			
DUNIX	*	*			
Eden	*	*	*		*
Freedomnet	*	*			
Gothic	+	+	+	+	*
Guide	*	*		*	
Gutenberg	*	*	*	*	*
HERMIX	*	*			
JASMIN	*				
LOCUS	*	*	*	*	*
MACH	*	*			*
MARUTI	*	*			
MOS	*	*			
Newcastle		*			
NEXUS	*	*	+	+	*
PEACE	*	*	*		
Profemo	*	*		*	
Prospero	*	*			
PULSE	*	*	*		
QuickSilver	*	*			*
RHODOS	*	*	(*)		(*)
Saguaro	*	*	+		
SOS	*	*			*
Sprite	*	*		*	
V	*	*			
WANDA	*	*			
Wisdom	*	+			

Table 7.6: Distributed Operating Systems — Part 1

Name	OS	Heterogeneity CPUs
Accent		PERQ
Alpha		680X0, MIPS-based
Amoeba	(UNIX)	VAXs, SUN 3, (386s, SPARC)
Argus	Ultrix	MVAXs
BirliX	UNIX 4.3 BSD	SUN 3
Cambridge	TRIPOS, (UNIX)	LSI-4, 68000, Z-80, PDP-11/45, VAXs, SUNs
Charlotte	(UNIX)	Crystal multicomputer, VAX-11/750
CHORUS	UNIX, SCO SVR3.2	multi-vendor HW
Clouds	UNIX	VAX-11/750, SUN 3
Cosmos	UNIX	
Cronus	UNIX V7, 4.2 BSD, VMS	68000, VAXs, SUNs, BBN C70
DIOS	UNIX V, AOS/VS	MV8000, SMD, VAXs
DACNOS	VM/CMS, PC DOS, VMS	VAXs, IBM PC, IBM/370
DEMOS/MP		Z8000
DUNE	UNIX V, some BSD	68000
DUNIX	UNIX 4.x BSD	VAXs
Eden	UNIX 4.2 BSD	SUN, VAXs
Freedomnet	UNIX	multi-vendor HW
Gothic		BULL SPS7 MP
Guide	UNIX (MACH)	Bull DPX, DECstation3100, SUNs
Gutenberg	Ultrix, Dynix	VAXstations, Sequent Symmetry
HERMIX	UNIX	68000
JASMIN	UNIX V	
LOCUS	UNIX 4.x BSD, V	DEC VAX/750, PDP-11/45, IBM PC
MACH	UNIX 4.3 BSD	VAXs, SUN 3, NS32032, MacII, I386
MARUTI	UNIX (MACH)	SUN 3, DEC 3100, SPARC/VAXstations
MOS	UNIX V7	PDP-11, PCS/CADMUS 9000
Newcastle	UNIXs	multi-vendor HW
NEXUS	UNIX 4.2 BSD	SUNs
PEACE	(UNIX)	SUPRENUM, 680X0
Profemo	UNIX 4.2 BSD	DEC VAXs
Prospero	UNIX	
PULSE		LSI-11/23
QuickSilver	(VM/CMS)	IBM RTs, IBM 370, RS/6000
RHODOS		SUN 3/50
Saguaro	UNIX	SUN
SOS	UNIX (SunOS)	SUNs
Sprite	UNIX 4.3 BSD	SUNs, DEC&SPARCstations, Sequent Symmetry
V	UNIX 4.x BSD	SUN 2/3, VAXstationII
WANDA	UNIX	VAXs, Firefly, Acorn ARM3s
Wisdom		T414, T800, 400+ transputer, SUN

Table 7.7: Distributed Operating Systems — Part 2

| Name | Changes made | | Communication | | shared memory | RPC-based | VC | Connection | |
	new kernel	new layer	standard protocols	specialized protocols				datagram	pipes/streams
Accent	*	*		*				*	
Alpha	*		XTP	+				*	
Amoeba	*			Amoeba		*			
Argus		*		ACP/IP				*	
BirliX	*	(*)			*	*		*	
Cambridge	*		UDP/IP	BBP, SSP, BSP		*		*	*
Charlotte	*		*				*	*	
CHORUS	*	*	*		+	*		*	
Clouds	*			DSM	*	*			
Cosmos	*								
Cronus	*		TCP, UDP/IP	VLN		*	*	*	
DIOS	(+)	*				+		*	
DACNOS		*	OSI TLI			*		*	
DEMOS/MP	*			*		*		*	
DUNE	*		TCP/IP	*	*	*	*	*	
DUNIX	*		TCP/IP	*				*	
Eden		*		Accent		*			
Freedomnet		*		RSC		+			
Gothic	*		TCP/IP			*			
Guide	*	(*)			*				
Gutenberg		*	TCP/IP			*			
HERMIX	*					*			*
JASMIN	*			Paths					
LOCUS	*			*			*		
MACH	*			*	*	*		*	
MARUTI		*	*						
MOS	*		UDP/IP			*		*	
Newcastle	*	*	UDP/IP,LLC1	BBP		*		*	
NEXUS	*		UDP/IP			*		*	
PEACE	*		UDP/IP	*		*		*	
Profemo		*	UDP/IP		*	*			*
Prospero		*	UDP/IP					*	
PULSE	*					*	*		
QuickSilver	*			*		*		*	
RHODOS	*			RRDP/IP		*		*	
Saguaro									*
SOS	*		*	*		*			
Sprite	*			*		*			
V	*			VMTP		*		*	
WANDA	*			MSNL/MSDL		*			
Wisdom	*					*		*	

Table 7.8: Distributed Operating Systems — Part 3

Name	Semantics			Naming		Security			
	may be	at most once	exactly once	object-oriented	hier-archical	en-cryption	special HW	capa-bilities	mutual auth.
Accent		*			*			*	
Alpha		*	*	*				*	
Amoeba		*		*		*	*	*	
Argus			*	*					
BirliX	*			*	(*)	*		+	*
Cambridge		*	+	(*)	*			*	*
Charlotte	*				*			*	
CHORUS	*		(*)	*				*	
Clouds			*	*					
Cosmos		*		*				+	
Cronus	*			*				+	
DIOS	*				*				
DACNOS		*			*				*
DEMOS/MP	*		*		*			+	
DUNE			*		*				
DUNIX		*			*				
Eden		*		*				*	
Freedomnet		*			*	(*)			*
Gothic			*	*	(+)	·			
Guide			*	*					
Gutenberg			*	*				*	
HERMIX					*				*
JASMIN	*				*			*	
LOCUS		*	*		*				
MACH		*			*			*	
MARUTI			*	*		*		*	
MOS	*				*				
Newcastle		*	*		*				
NEXUS		*	*	*					
PEACE	*	*	*	*	(+)				
Profemo				*				*	
Prospero	*			*	(+)				*
PULSE	*				*				
QuickSilver		*			*				
RHODOS		*	*	(+)	*	*		*	*
Saguaro					*				
SOS	*	*		*				+	
Sprite		*			*				
V		*			*				
WANDA					*				(*)
Wisdom		*			*				

Table 7.9: Distributed Operating Systems — Part 4

Name	Availability				Failures				Object/Process Mobility
	synchro-nization	TA	nested TA	repli-cation	recovery client crash	recovery server crash	stable storage	orphan detection	
Accent					+	+			*
Alpha	*	*	*	*	*	*	*	*	*
Amoeba	*					*	*	*	
Argus	*	*	*		*	*	*	*	
BirliX	*			(*)	*	*			(*)
Cambridge		*			+	*		+	
Charlotte						*			*
CHORUS	*	(*)		*					+
Clouds	*	*	*		*	*	*		
Cosmos	*	*		*					
Cronus				*					*
DIOS	*								
DACNOS	*								
DEMOS/MP					*	*			*
DUNE				+	*				*
DUNIX									
Eden	*	*	*	*		*			*
Freedomnet					*		+	*	+
Gothic	*				*	*	*		
Guide	+	*	*						
Gutenberg	*	*		*			*		
HERMIX	*	*							
JASMIN									
LOCUS	*	*	*	*	*	*			*
MACH					+	+			*
MARUTI	*			*					
MOS				*					*
Newcastle	*				*	*		*	
NEXUS	+	*	*	+					
PEACE				*					
Profemo	*	*	*						
Prospero	*								*
PULSE	*			*					
QuickSilver		*				*	*		
RHODOS				(*)					*
Saguaro				*					
SOS					+	*			*
Sprite	*			+		*			*
V	*	(*)		(*)					*
WANDA		(*)					(*)		
Wisdom								'	*

Table 7.10: Distributed Operating Systems — Part 5

Name	Transparency				
	location	access	replication	concurrency	failure
Athena	+	+			
Avalon	*	*			*
DAPHNE	*	*			
DASH		*			
Emerald	*	*	+		
Enchère		*	*	*	*
Galaxy	(*)	(*)	(*)	(*)	(*)
GAFFES	*	*	*	*	
Grapevine	*	*	*		
HCS	*	*			
INCAS	+	+		+	
ISIS	+	+	+	+	+
Medusa	*	+			
Meglos	+	+			
Mirage	*	*	+	+	
NCA/NCS	*	*	*	*	*
Plan 9	*				
Psyche	*	*			
SMB		*			
Symunix	*	*			
Synthesis	*	*	+	+	+
x-Kernel	+	+			

Table 7.11: Closely Related Systems — Part 1

Name	Heterogeneity	
	OS	CPUs
Athena	UNIX	multi-vendor HW
Avalon	MACH	VAXs, SUNs, IBM RTs, PMAXs
DAPHNE	UNIX, (VMS, MS-DOS)	SUN 3/SPARC, VAXstation, IBM PC
DASH	(UNIX)	SUN 3/SPARC
Emerald	UNIX, Ultrix	VAXs, SUN 3
Enchère		Intel 8086/8085
Galaxy	Acis 4.2	IBM RT PCs
GAFFES	(all kind)	(all kind)
Grapevine	Pilot	Alto, 8000 NS
HCS	UNIX, XEROX XDE	VAXs, SUNs, Tektronix, XEROX D
INCAS		68000, SUN
ISIS	UNIX (VMS, VM/CMS)	multi-vendor HW
Medusa	(StarOS)	Cm*
Meglos	UNIX	68000, VAXs, PM-68k
Mirage	UNIX V	VAX 11/750 (IBM PS 2)
NCA/NCS	DOMAIN/IX, UNIX 4.x BSD, VMS, MS/DOS	SUNs, IBM PCs, VAXs
Plan 9	UNIX	Silicon GPS (MIPS 6280), Gnot
Psyche		BBN Butterfly, GP1000 line
SMB	UNIX V, UNIX 4.x BSD, VMS, DOS	multi-vendor PC
Symunix	(UNIX V7) UNIX 4.3 BSD	NYU Ultracomputer, IBM RP3
Synthesis	(UNIX, MACH)	68020, 68030
x-Kernel		SUN 3 (MIPS, SPARC RISC)

Table 7.12: Closely Related Systems — Part 2

| Name | Changes made | | Communication | | | RPC-based | Connection | | |
	new kernel	new layer	standard protocols	specialized protocols	shared memory		VC	datagram	pipes/ streams
Athena		+	TCP/IP				*		
Avalon		*		Camelot		*			
DAPHNE		*	UDP/IP			*		*	
DASH	+		TCP/IP	SRP	(*)				*
Emerald		*							
Enchère	*							*	
Galaxy	*			*	(*)				
GAFFES		+	*			*	*		
Grapevine				PUP				*	*
HCS		*	*			*			
INCAS	*		TCP/IP				*		*
ISIS		*	UDP/IP			+		*	
Medusa	*				*				*
Meglos	*			*			*		
Mirage	*			*	*				
NCA/NCS		*	UDP/IP			*	*	*	
Plan 9	*			Datakit	*				
Psyche	*		UDP/IP		*	+		*	
SMB	*			SMB			*		
Symunix	*		4.3 BSD		*				*
Synthesis	*		TCP/IP	*			*	(+)	
x-Kernel	*		TCP/IP	*		*	*	*	

Table 7.13: Closely Related Systems — Part 3

Name	Semantics			Naming		Security			
	may be	at most once	exactly once	object-oriented	hier-archical	en-cryption	special HW	capa-bilities	mutual auth.
Athena	*				*	*			*
Avalon			*	*					
DAPHNE		*			*				*
DASH					*	(*)			(*)
Emerald	*			*					
Enchère			*					*	
Galaxy				*				+	
GAFFES			*		*	*	*	*	*
Grapevine					*				
HCS									
INCAS				*					
ISIS					*	*	+		
Medusa				*					
Meglos		*			*				
Mirage					*				
NCA/NCS		*		*					
Plan 9					*				*
Psyche				*					
SMB	*				*				
Symunix	*				*				
Synthesis	+	+	+	+		(*)	*	+	(*)
x-Kernel									

Table 7.14: Closely Related Systems — Part 4

Name	Availability				Failures				Object/ Process Mobility
	synchro-nization	TA	nested TA	repli-cation	recovery client crash	recovery server crash	stable storage	orphan detection	
Athena									
Avalon	*	*	*	+	*	*	*		
DAPHNE								*	
DASH									
Emerald	+			+					*
Enchère	*	*	*				*		*
Galaxy				(*)					(*)
GAFFES	*	*	*	*					
Grapevine				*					
HCS									
INCAS	*								
ISIS	*	+		+	*	*	+		
Medusa	*								*
Meglos	+								
Mirage	*			+					+
NCA/NCS	+			*					
Plan 9									
Psyche				(*)					(*)
SMB									
Symunix	*								*
Synthesis	*	+		+					
x-Kernel									

Table 7.15: Closely Related Systems — Part 5

Chapter 8

Related Projects

In this chapter I present a list of related projects which are not described in the survey. The list contains a short description of and at least a main reference to each project.

8.1 Acorn

The Acorn File Server developed at Acorn Computers plc supports diskless workstations in schools by providing a shared hierarchical file system. The Acorn File Server ran on M 6502 with floppy disks.

REFERENCE: [556]

8.2 Agora

The main goal of the Agora approach is to facilitate heterogeneous parallel programming. It supports the operating system level as well as the programming environment. Moreover, it provides additional mechanisms for process interaction in distributed systems. The Agora system layer allows processes to interact through events and a special kind of memory sharing with emphasis on optimizing static relationships between processes. Thereby it provides the illusion of memory sharing through software support. It is implemented on top of MACH (Sect. 4.13).

Agora runs on DEC VAXes, IBM RT PC, SUNs, and Encore Multimax. The languages supported are C, C++, and CommonLisp.

CONTACT: Robert Bisiani, Dept. of Computer Science, Carnegie Mellon Univ., Pittsburgh PA 15213, USA

REFERENCES: [557], [558]

8.3 Amber

Amber is a follow-on of the Emerald language (Sect. 6.5). Amber provides for distributed parallel programs using a network of multiprocessors. It has lightweight threads and Emerald-style object mobility. Amber is written in C++.

CONTACT: Hank Levy, Dept. of Computer Science and Engineering, FR-35, Univ. of Washington, Seattle WA 98195, USA
e-mail: levy@cs.washington.edu
    REFERENCE: [559]

8.4   Arachne

Arachne is the direct predecessor of the Charlotte operating system (the Crystal project, Sect. 4.3). Arachne is a distributed operating system kernel that was designed at the Computer Sciences Department of the University of Wisconsin-Madison.
    CONTACT: Raphael A. Finkel, Dept. of Computer Science, Univ. of Kentucky, Lexington KY 40506, USA
    REFERENCES: [560], [561]

8.5   Arca

Arca is a centralized file system server developed at the Department of Computing of the University of Lancaster. Arca is based on the Cambridge File Server (Sect. 4.2) but, unlike CFS, which runs on an operating system kernel called TRIPOS, Arca is written as a monolithic program on a bare machine.
    Arca serves two different LANs: an Ethernet-type network called *Strathnet*, and a Cambridge Ring-type network called *Polynet*. VAX-11/750s, PDP-11s, and M68000 hosts are attached to these networks. Two access protocols are being used by Arca: the Single-Shot Protocol and the RPC. An atomic update mechanism is implemented. Ports, represented by a 16-bit integer, are known addresses to which requests and replies can be sent.
    In future, it is planned to integrate stable storage into Arca, and to distribute the file server among two or more machines.
    CONTACT: S. Muir or D. Hutchinson or D. Shepherd, Dept. of Computing, Univ. of Lancaster, Bailrigg, Lancaster LA1 4YR, UK
    REFERENCE: [562]

8.6   Arcade

Arcade is a distributed operating system kernel with the objective to build a system which executes on distributed and potentially heterogeneous processors. It supports two primary abstractions: (active) tasks and (passive) data units.
    The Arcade kernel provides creation/destruction of tasks and data units, task migration, distributed shared data units, data unit locking and update propagation, and a task synchronization facility.
    STATUS: A prototype is currently operational and runs on a collection of 80386-based PS/2s connected by a token ring network. A port to System/370 processors is in progress.

CONTACT: David Cohn, Dept. of Computer Science and Engineering, Univ. of Notre Dame, Notre Dame IN 46556, USA

e-mail: dcl@cse.nd.edu

or Bill Delaney, Dept. of Electrical and Computer and Engineering, Univ. of Arizona, Tucson AZ 85721, USA

e-mail: wpd@ece.arizona.edu

REFERENCES: [563] – [566]

8.7   Archons (ArchOS)

The Archons project to create new paradigms for real-time distributed computing took place at Carnegie Mellon University's Computer Science Department during 1979 through 1987. It was an unusually large academic project comprised primarily a conceptual and theoretical element (e.g., system principles and architectures, distributed transactions, distributed decision making and control, real-time scheduling), computer architecture studies (e.g., RISC/CISC ISAs, special-purpose algorithmic accelerators), and two real-time distributed operating system efforts: Alpha (Sect. 5.1), which became the centerpiece of the Archons project in 1985, and a new project of its own in 1987, and ArchOS, which was terminated in favor of Alpha.

The novel direction and approaches of the Archons project attracted the participation of numerous researchers from industry and other universities, as well as from within CMU. Consequently, the literature shows that Archons directly and indirectly played a unique role in setting an agenda for academic and industrial research in real-time computing — in both theoretical and system contexts — that continues to thrive today.

The Archons project was funded by a large number of DoD agencies and industrial corporations.

CONTACT: E. Douglas Jensen, Concurrent Computer Corporation, One Technology Way, Westford MA 01886, USA

e-mail: jensen@westford.ccur.com

REFERENCES: [567] – [571]

8.8   Argos

Argos is a message-based operating system for the *Global Memory Message Passing* (Gmmp) multiprocessor architecture. A Gmmp architecture optimizes the message passing model of computation, in which the process access environments are strictly isolated.

Argos provides each process a private address space. Communication and synchronization are performed via messages. On the other hand, global memory is used to implement rapid message passing and copy-on-write between processes.

STATUS: Currently, the designers are building a machine of the Gmmp-class called the *Virtual Port Memory* (VPM) machine. However, hardware and operating system are not yet developed.

CONTACT: Eric E. Johnson, Dept. of Electrical and Computer Engineering, New

Mexico State Univ., Box 3-O, Las Cruces NM 88003, USA
e-mail: ejohnson@nmsu.edu
    REFERENCES: [572] – [574]

8.9   Arjuna

Arjuna is an object-based fault-tolerant distributed programming system. It is being
designed and implemented in C++ on a set of UNIX workstations connected by an Ethernet. Arjuna employs atomic actions (transactions) for structuring programs. Programs
operate on objects which are instances of abstract data types. In Arjuna, objects are
long-lived, persistent entities and are the main repositories for holding system state. By
ensuring that objects are only manipulated within atomic actions, it can be guaranteed
that the integrity of the objects – and hence the integrity of the system – is maintained
in the presence of failures such as node crashes and the loss of messages.

There are several aspects of Arjuna that are novel. These are for example: Use of
multicast RPC for managing a variety of activities of process groups, such as commit
processing, use of type-inheritance for specifying and implementing recovery and concurrency control strategies, and use of nested, concurrent actions for program structuring.
Arjuna objects are typically resident in local object stores in a passive mode. An object
invocation activates the stored object. When an object commits (terminates), active
objects are put back on the respective object stores (passivated).

STATUS: Major parts of the Arjuna system have been implemented. Currently, the
designers are integrating these parts to form a complete system.

CONTACT: Graham Parrington, Computing Laboratory, Univ. of Newcastle upon
Tyne, Claremont Tower, Claremont Road, Newcastle upon Tyne NE1 7RU, UK
e-mail: graham@nss.cs.ucl.ac.uk
    REFERENCES: [575] – [577]

8.10   Boston Community Information System (BCIS)

The Boston Community Information System (BCIS) has the main goal of exploring
a type of distributed system architecture that can provide sophisticated information
services to a million people, for example electronic mail, name server information, and
distributed database inquiries.

BCIS is a *polychannel* approach integrating the advantages of both, the duplex (such
as R*, Sect. 8.51) and simplex systems (such as Teletex systems). For security reasons
all broadcasted data is cryptographically protected through a private key encryption.

The BCIS project began during late 1982 and early 1983. A prototype version is now
in use at a test population of about 200 users.

CONTACT: David K. Gifford, Laboratory for Computer Science, MIT, 545 Technology Square, Cambridge MA 02139, USA
e-mail: gifford@brokaw.lcs.mit.edu
    REFERENCE: [578]

8.11   Camelot

In the Camelot project, a distributed (nested) transaction facility combined with high performance has been constructed. It executes on a variety of uni- and multi-processors on top of MACH whose description can be found in the survey (Sect. 4.13).

For atomicity reasons, Camelot is used in systems such as Avalon (Sect. 6.2) and Coda (Sect. 2.4).

CONTACT: Dan Potapshyn, Mt. Xinu, 2560 Ninth Street, Berkeley CA 94710, USA
e-mail: dan@mtxinu.com
REFERENCES: [579] – [582]

8.12   Carnegie Mellon Central File System (CMCFS)

The Carnegie Mellon Central File System (CMCFS) was designed in early 1980. It provides two kinds of update schemes. First, the immutable file approach by which a new version is created with each update, and second, a traditional update based on transactions. CMCFS is part of the Spice project.

REFERENCE: [583]

8.13   Choices

Choices is a family of operating systems developed in the Department of Computer Science at the University of Illinois, Urbana. Each operating system is built using an object-oriented approach. They export conventional services to the applications.

REFERENCE: [584]

8.14   Circus

Circus is a replicated remote procedure call facility. The so-called *paired message protocol* of Circus is based on the connectionless DARPA UDP. It is responsible for the segmentation of messages which are longer than a single datagram. Multiple segments are allowed to be sent out before one has been acknowledged. It is currently implemented in user code under UNIX 4.2 BSD.

REFERENCES: [585], [586]

8.15   Clearinghouse

The commercial product Clearinghouse is a distributed and replicated name server. It is developed from Xerox Corp. and is a successor to Grapevine (Sect. 6.9). However, it provides a hierarchical name space with three levels and uses different replication algorithms. The three levels of the Clearinghouse name service consist of organization, domain, and object.

Currently, about 350 Clearinghouse services are in operation all over the world. The main problem resulting from this size is the long propagation of updates (days or weeks) and the heavy Clearinghouse-induced internetwork traffic.

REFERENCE: [587]

8.16   Cocanet

Cocanet is a local computer network based on the UNIX operating system. It extends the existing file system name space by adding the host name to the resource name, for example /vax1/usr/bin. Additionally, standard creation primitives can be used to execute remote programs.

REFERENCE: [588]

8.17   Computing System for Societies of Agents (CSSA)

The Computing System for Societies of Agents (CSSA) is an experimental distributed programming language. It is developed in the INCAS project (INCremental Architecture for distributed Systems) at the University of Kaiserslautern.

CSSA is a high level programming language for expressing distributed algorithms involving many cooperating tasks. It is based on the notion of *actors*. Actors are active objects that communicate with each other solely by asynchronous message passing. Its object-oriented message passing philosophy, called "module-message paradigm," is especially well suited to distributed computing. Several applications, including a distributed calendar and appointment system and a distributed discrete event simulation system, were written in CSSA.

STATUS: A preliminary LISP-based sequential version of CSSA was finished in 1979. In 1983 a concurrent implementation of the compiler and the CSSA system was completed. The compiler was written in SIMULA and ran on IBM and Siemens mainframes. It translated CSSA programs into SIMULA code. To gain portability, the compiler has been rewritten in C generating low-level C-code for a virtual stack-machine. The distributed runtime system of CSSA is formed by a set of communicating virtual machines. Since 1985 a prototype version has been available. In 1986 CSSA has been ported to a distributed UNIX 4.2 BSD system. It is now available on a network of UNIX machines and SUN workstations connected by Ethernet using a TCP/IP-based protocol. In 1990 an implementation on a Transputer cluster was started.

CONTACT: Horst Mehl, Fachbereich Informatik, Univ. Kaiserslautern, Postfach 3049, D-6750 Kaiserlautern, Germany
e-mail: horst@informatik.uni-kl.de

REFERENCES: [589], [590]

8.18 CONIC

The CONIC environment is designed to support the construction and operation of software for distributed embedded systems. It employs a host/target approach, providing a comprehensive set of tools for program compilation, building, debugging and execution on the host, and supports distributed operation on the targets. It is being developed at the Dept. of Computing, Imperial College, London, UK.

REFERENCE: [591]

8.19 Customer Information Control System (CICS)

The Customer Information Control System (CICS) is a commercial product of the IBM Corp. designed as a transaction processing system. CICS provides transaction services to resource managers such as DL/1, System 2000, and System R. It also provides a record interface to terminals and to sessions through virtual circuits. CICS is access transparent using a classic RPC facility. The RPC works for queues, processes, files, DL/1 databases, System 2000 databases, and other objects. CICS maintains a system-wide log for transactions and uses the standard two-phase commit protocol to coordinate the transaction commit and to deal with node or link failures. Replication transparency is not supported. Global deadlocks are detected by timeouts.

CICS implements IBM's SNA (System Network Architecture) and provides communication with non-CICS and non-IBM systems via a well defined interface.

REFERENCE: [592]

8.20 Datacomputer

The Datacomputer is a data management and storage utility developed to share resources in the Arpanet. Clients are normal Arpanet hosts. The Datacomputer was implemented on DEC PDP-10 and offered services on the Arpanet in late 1973.

REFERENCE: [593]

8.21 Deceit

Deceit is a fault-tolerant file system that uses the NFS protocols (Sect. 2.8) to communicate with servers and clients but uses ISIS internally (Sect. 6.12). File replication is varied based on performance considerations and availability requirements.

CONTACT: Ken Birman, Dept. of Computer Science, Cornell Univ., 4105 Upson Hall, Ithaca NY 14853, USA
e-mail: birman@cs.cornell.edu or isis@cs.cornell.edu

REFERENCE: [594]

8.22   DEMOS

DEMOS is an operating system for a Cray 1. Many of the facilities are being used in
the distributed version DEMOS/MP which is described in the survey (Sect. 4.4).
REFERENCE: [595]

8.23   DFS 925

DFS 925 is a Distributed File System for the workstations 925. DFS 925 has been de-
veloped with the goal of building an integrated system providing location transparency
and atomic transactions. Replication of files is not supported.
CONTACT: Office Systems Laboratory, IBM San Jose Research Laboratory, San Jose
CA 95120-6099, USA
REFERENCE: [596]

8.24   DISTRIX

DISTRIX is a message-based implementation of UNIX System V for a network of workstations developed by Convergent Technologies, USA. The kernel is a collection of server
processes based on a real-time operating system called the *Foundation System.* Processes
communicate either synchronously or asynchronously in a location-transparent fashion.
REFERENCE: [597]

8.25   Dragon Slayer

Dragon Slayer is a distributed operating system developed at the Wayne State University.
A distributed resource scheduling algorithm implemented makes Dragon Slayer a special
development with provision for services for novel applications, e.g., in office information
systems.
CONTACT: Computer Science Dept., Wayne State University, Detroit MI 48202, USA
REFERENCE: [598]

8.26   Echo

Echo is a distributed file system which is currently being implemented at the Digital
Equipment Corporation. Its main goal is to explore issues of scaling, availability, and
performance.
Echo provides a global, hierarchical name space. The name space is structured as a
collection of subtrees, called *Echo Volumes.* It supports UNIX file system semantics as
an option.
To increase the file access performance and to reduce disk seeks, Echo uses distributed
caching on clients as well as a logging scheme on the file server. Using the logs that record
information about updates that are in progress, consistency-preserving crash recovery is
provided.

To increase the file availability, Echo uses a primary-site block-level replication scheme. This scheme maintains the primary's quorum using keep-alives. Writes are performed at all sites available, i.e., up and in communication. Reads are only performed at a single site, since there is at most one primary that handles updates. In case the primary site fails, a re-election strategy is used.

STATUS: Echo is beginning to turn over. Replication and failover are already coded and just tested. Future research includes the increase of the failover-performance.

CONTACT: Timothy Mann, or Andy Hisgen, Digital Equipment Corporation, Systems Research Center, 130 Lytton Ave., Palo Alto CA 94301, USA
e-mail: {mann,hisgen}@gatekeeper.dec.com

REFERENCES: [599] – [601]

8.27   Encompass

Encompass is a distributed database management system for the TANDEM computers. It is built on top of a recovery mechanism based on replication of both hardware and software.

REFERENCE: [602]

8.28   Felix

Felix is a file server for the Helix distributed system (Sect. 3.2) developed at Bell-Northern Research. It supports virtual memory and sharing of data and provides secure access by the use of capabilities.

CONTACT: Marek Fridrich, Bell-Northern Research, PO Box 7277, Mountain View CA 94039, USA

REFERENCE: [603]

8.29   Ficus

Ficus is a replicated distributed file system for UNIX currently being developed at UCLA. Its main goal is to scale to very large, e.g., nationwide, networks.

Ficus is a new "stackable" layer that can be added to existing operating systems providing a *vnode* interface. To increase the file access availability as well as at least the read file access performance, a collection of file volume replicas is set up at various nodes in a replication-tranparent manner. A local replica is accessed via procedure calls (pure vnode operations). To access remote replicas, the current Ficus implementation uses NFS (Sect. 2.8) to map the vnode operations across the network. Using NFS makes Ficus available to all machines where a NFS client implementation exists.

Ficus implements an optimistic "one copy availability", i.e., conflicting updates are detected and reported to a higher level. User interaction may be required to reinstall a consistent view to the replicas.

STATUS: Operational and native environment for the developers. Testing outside UCLA is expected 1991.

CONTACT: Gerald Popek, Thomas Page, or Richard Guy, Dept. of Computer Science, Univ. of California, Los Angeles CA 90024-1596, USA
e-mail: {popek,page,guy}@cs.ucla.edu

REFERENCES: [604] – [610]

8.30   FileNet

FileNet is a distributed file system developed at FileNet Corporation as part of a commercial product. Furthermore, it is a distributed application system supporting document image processing. The main goal of the FileNet file system development was to have specific support for the so-called *read-mostly* workload imposed by this kind of application.

FileNet runs on M 68000-based workstations and servers. Other hardware-type systems can interact with the FileNet servers at the application software level. The UNIX file system semantics are provided, with one exception concerning the file access times stored at the inode-level.

STATUS: The FileNet design began in late 1982, with the first versions implemented in 1983 and 1984 The early version relied on the concepts of LOCUS (Sect. 4.12). The overall objective of the new versions was the minimization of the server workload in terms of decreasing their responsibilities.

CONTACT: David A. Edwards and Martin S. McKendry, FileNet Corporation, 3530 Hyland Avenue, Costa Mesa CA 92626, USA

REFERENCE: [611], [612]

8.31   Firefly

Firefly is a shared-memory multiprocessor workstation of DEC Systems Research Center that contains from one to seven processors. The processors have coherent caches, so that each sees a consistent view of the main memory. See the Topaz reference (Sect. 8.65) for a description of the software system.

The Firefly RPC is the primary communication mechanism between address spaces, both for inter-machine and intra-machine communication. The performance is shown to be good without sacrificing any functionality. The packet exchange protocol is built on top of UDP/IP.

CONTACT: Michael D. Schroeder, Digital Equipment Corporation, Systems Research Center, 130 Lytton Avenue, Palo Alto CA 94301, USA
e-mail: mds@gatekeeper.pa.dec.com

REFERENCES: [613], [614]

8.32   Generic File System (GFS)

The Generic File System for UNIX (GFS) has been created by Digital Equipment Corporation to support all types of hierarchical file systems (local, remote, stateful, stateless),
for example NFS (Sect. 2.8), Berkeley FFS or MS-DOS file system. GFS controls all
common file system resources and becomes the sole interface for file system operations.

REFERENCE: [615]

8.33   Helios

Helios is a distributed operating system for transputer systems. The concept of Helios was influenced by the Cambridge Distributed Computing System (Sect. 4.2) and
Amoeba (Sect. 5.2). Furthermore, the user interface and the system call mechanisms
were influenced by UNIX.

Currently Helios runs on a host, such as a SUN workstation or an IBM PC, and a
transputer network. Each transputer runs the Helios nucleus which is structured in a
client/server way. The host processor provides the I/O-facilities.

Clients send messages to servers in order to get the services of the operating system.
Thereby, messages are sent through ports using a stateless General Server Protocol.
Ports are logical end-to-end connections. The physical link-connections of the processors
are managed by Helios exclusively.

Objects in the system, such as files, devices, tasks, etc., are protected by encrypted
capabilities.

STATUS: Active. Atari announced a transputer-workstation running Helios.

REFERENCES: [616] – [618]

8.34   HERON

HERON is a successor to DAPHNE (Sect. 6.3). The DAPHNE approach supports the
development of distributed application systems for heterogeneous environments without
using special "distributed" programming languages and without modifications of the existing system software.

In comparison with DAPHNE, HERON will support non-sequential, object-oriented systems. As reference language, a non-sequential extension of Eiffel will be used.

CONTACT: Klaus-Peter Löhr or Lutz Nentwig, Freie Univ. Berlin, Fachbereich
Mathematik, Inst. für Informatik, Nestorstr. 8–9, D-1000 Berlin 31, Germany
e-mail: lohr@fubinf.uucp or nentwig@fubinf.uucp

REFERENCE: [619]

8.35   Intelligent Distributed Resource Processing System (IDRPS)

The Intelligent Distributed Resource Processing System (IDRPS) of the Software Productivity Consortium should provide transparent access to a wide variety of resources without any concern for access techniques. The system will utilize as many components of existing systems as possible. NFS (Sect. 2.8) is chosen to provide remote file access. Transparent remote process execution can be achieved by adding a layer of software on top of NCS (Sect. 6.16). Some features like authentication and load balancing should be added. The environment includes Apollo, VAX, Gould, TI Explorers, SUN and Symbolic AI machines.

REFERENCE: [620]

8.36   Interim File System (IFS)

The Interim File System (IFS) developed at Xerox PARC is a distributed file system organized in a tree of directories. IFS has been operational since 1975 in connection with Alto microcomputers. It is written in the BCPL language and "talks" a protocol built for it: PUP FTP.
IFS provides a repository of shared files and is used for backup and archiving of private files.

STATUS: IFS is still in operation at XEROX.
REFERENCE: [621]

8.37   Language for Distributed Systems (LADY)

The object-oriented Language for Distributed Systems (LADY) was developed for the implementation of distributed operating systems. It provides broadcast and multicast communication modes at the language level. Furthermore, it allows for the modification of communication paths, and the creation and deletion of *teams* at runtime. A team is an autonomous entity and implements a set of related operating system functions. Data exchange among teams (inter-team communication) is exclusively performed by message passing using so-called export and import interfaces. Within the teams, processes and monitors are the most important building blocks. Communication among processes (intra-team communication) is accomplished via monitors.

The structuring concept of LADY are expressed by three language levels: (a) the system level, (b) the team level, and (c) the module level. Similar concepts can be found in Argus, CHORUS, and Eden's EPL (Sects. 5.3, 5.5, 5.9 resp.).

OTHER ISSUES: Applications written in LADY include a prototype version of a distributed database system for CAD/CAM, a distributed simulation system for VLSI-applications, and a small operating system.

STATUS: Research began in 1980. At the beginning of 1983 a first version of LADY was implemented and some experience was gained. Between 1985 and 1989 the complete

LADY programming environment was available and used in several distributed applications. The project now aims at the development of an integrated environment for general purpose distributed programming that supports the activities of design, implementation, execution, as well as behavior and performance analysis of distributed applications.

CONTACT: Jürgen Nehmer, Fachbereich Informatik, Univ. Kaiserslautern, Postfach 3049, D-6750 Kaiserlautern, Germany

e-mail: nehmer@informatik.uni-kl.de

REFERENCES: [622] – [625]

8.38   Lynx

Lynx is a message-passing language for distributed applications. It was developed at the University of Wisconsin to support both application and system software in a single conceptual framework. In 1984, Lynx was first implemented on the Charlotte operating system (Sect. 4.3). In 1986, at the University of Rochester, it has been ported to the BBN Butterfly multiprocessor and its Chrysalis operating system. Other designs exist for UNIXs using TCP/IP and for an experimental, paper-designed system called SODA.

Lynx supports topology changes as well as protection with *links*, a virtual circuit abstraction. Lynx subdivides each heavyweight process into lightweight threads of control as a program structuring tool. Threads use coroutine semantics: only one thread is running at a time, and is rescheduled when it blocks. RPC-style message passing is provided. A thread that initiates a RPC blocks until a reply is received. As mentioned before, other threads are allowed to run in the meantime, so the (main) heavyweight process is not blocked.

STATUS: An implementation of Lynx is being developed for the Psyche multiprocessor operating system (Sect. 6.18).

CONTACT: Michael S. Scott, Univ. of Rochester, Dept. of Computer Science, Rochester NY 14627, USA

e-mail: scott@cs.rochester.edu

REFERENCES: [626], [627]

8.39   Management of Distributed Systems (MANDIS)

The MANDIS project stands for MANagement of DIstributed Systems. MANDIS explores management issues in an environment of local and wide area networks. Thereby, it provides transparent communication in an internetwork and tools for managing a distributed system. A prototype which is based on Amoeba (Sect. 5.2) has been installed in four European countries.

CONTACT: David Holden, or Alwyn Landsford, Building 7–12, Harwell Laboratory, Didcot, Oxon. OX11 0RA, UK

REFERENCES: [628], [629]

8.40   Melampus

Melampus is a new object-oriented distributed operating system approach. Its vision is to provide a significantly more powerful, more usable, and more flexible computing environment than those currently in existence. Its goal is to provide a common framework for describing, creating, querying, and evolving all of the objects in the system. In particular, the designers believe that providing a comprehensive query facility is necessary so that users can locate objects and derive interesting new relationships among them.

In current systems, the lack of a common data model for all information inhibits integration of applications. Melampus' initial goal is to define a data model so powerful and easy-to-use that it becomes the basis for all programming activities within the system, from shells and user-interfaces to applications to systems. Following the initial design effort, it is planned to implement the model and to experiment with it in order to verify its strengths and discover its weaknesses. The experiments will include using the system for day-to-day activities as well as coding interesting applications. The system must support a community of real users, and it must be scalable, distributed, and multi-user.

A particular focus of its data modeling efforts is to address two major stumbling blocks in the use and maintenance of current computing systems: the fact that data is hard to find and the fact that systems evolve. Locating data is a problem at all levels of a computing system. From a programmer looking for an implementation of priority queues, to a user trying to locate an old news article, to a personnel administrator trying to find employees who earn more than their managers, searching is one of the most common (and often most frustrating) ways that humans use computers. Thus, one of the central themes of the Melampus data model is to provide associative access to all entities in the system. This includes entities created and owned by users (e.g., databases, mailboxes, etc.) as well as those managed by the system (e.g., disk queues, processes, etc.).

Like the problem of locating data, the problem of evolution occurs at all levels of the system. A software system itself evolves because of improvements in the implementation, changes in the stated requirements of the system, a desire to integrate existing applications, etc. Furthermore, a software system built in Melampus must be able to model entities that evolve. For example, a database system should be able to model individuals who migrate through many roles in a lifetime, e.g., student, employee, system-user, etc. A program development environment must be able to model software systems that evolve through different versions and configurations.

To support multiple users in Melampus, it is planned to provide fine-grained access control based on access control lists; logically, there is an access rights check on every method invocation. While the designers believe that such fine control is critical in many cases, they also acknowledge that access checking is not needed for the vast majority of object accesses in a running system. For example, temporary objects that never escape a process's stack frame do not need protection. A major challenge, then, is to provide a comprehensive access control mechanism when it is needed, while avoiding an overhead penalty when it is not.

A second interesting implementation problem concerns storage management. The problems with explicit object deletion are well-known, i.e., dangling references and the creation of garbage. Thus, many object-oriented systems are garbage-collected systems. However, since Melampus should be a fully distributed system, any garbage collection scheme must also be distributed. While there have been some efforts in this area, it is not yet clear how well any of them will scale. Moreover, garbage collection suffers from a problem that is called the "squirreling syndrome." That is, large numbers of objects are neither used nor garbage-collected because users squirrel away references to objects and then forget about them. The designers are currently considering a scheme involving expiration dates that would allow the system to reclaim expired objects, leaving a tombstone in its place.

CONTACT: Luis-Felipe Cabrera, Laura M. Haas, Allen W. Luniewski, Joel E. Richardson, Peter M. Schwarz, Jim W. Stamos, IBM Almaden Research Center, 650 Harry Road, San Jose CA 95120-6099, USA

e-mail: {cabrera,laura,luniew,joelr,schwarz,stamos}@ibm.com

REFERENCES: [630] – [632]

8.41 Meta

Meta is a high-level application of the ISIS system (Sect. 6.12). It is a subsystem for instrumenting a network application or environment. Anything that has a value that changes over time can be declared as a Meta sensor. Sensors are accessible system-wide through a database, and abstractions like fault-tolerant sensors or composed sensors (average, min, max, but with rigorous temporal semantics) are supported. Similarly, this layer of Meta supports actuators: ways of influencing the environment by setting values. *Lomita* is a language built using Meta. Using it, one can easily specify algorithms that monitor and react to sensor conditions. Lomita includes a fault-tolerant interpreter that pushes evaluation as close to the sensors as possible for maximum performance.

CONTACT: Ken Birman, Dept. of Computer Science, Cornell Univ., 4105 Upson Hall, Ithaca NY 14853, USA

e-mail: birman@cs.cornell.edu or isis@cs.cornell.edu

REFERENCE: [633]

8.42 MICROS

MICROS is the distributed operating system for the MICRONET network computer. It is intended for control of network computers of thousands of nodes. A main design goal is the optimization of network costs, such as internode communication and processing throughput. MICRONET is a network of 16 loosely-coupled LSI–11 microcomputer nodes, connected by packet-switching interfaces to pairs of medium-speed shared communication buses (500 kB/second). Each node runs a private copy of the MICROS kernel written in Concurrent PASCAL. All network processes communicate via a uniform message passing system. For large networks, MICROS resources are managed in nested pools. Each management node controls all resources within its subtree of the network hierarchy.

Addressable entities have globally unique names, each consisting of three fields informing of its type, the creating node, and a unique number within that node.

STATUS: MICROS was intendeded to provide a viable system for computer science research and possible commercial development. The current status is unknown.

REFERENCE: [634]

8.43   MODOS

MODOS is a joint project of the RWTH Aachen and the Telenorma GmbH in Frankfurt, Germany. It is a distributed real-time operating system designed in a client/server relationship.

The MODOS-kernel design has been influenced by the V-kernel and PEACE approaches (Sects. 4.21, 5.15, resp.). It provides location-transparent, message-oriented interprocess communication as well as light-weighted processes, teams, and multitasking. Additional communication processes make it possible to use MODOS as a distributed system.

STATUS: Active.

CONTACT: Wolfgang Kubalski, Rogowski-Institut RWTH Aachen, Schinkelstr. 2, D-5100 Aachen, Germany, or Klaus Autenrieth, Telenorma GmbH, Zentrale Entwicklung, Postfach 10 21 60, D-6000 Frankfurt 1, Germany

REFERENCE: [635], [636]

8.44   Munin

Munin is a distributed shared memory system. It allows parallel programs written for shared memory multiprocessors to be executed on distributed memory multiprocessors. It provides a suite of consistency protocols, and implements release consistency (entirely in software). Thus, network latency can be masked.

Munin runs on an Ethernet network of 16 SUN 3 processors. A port to MACH (Sect. 4.13) is planned.

CONTACT: Willy Zwaenepoel, Dept. of Computer Science, Rice Univ., USA
e-mail: willy@rice.edu

REFERENCES: [637], [638]

8.45   Network Workstations (NEST)

The NEST project (NEtwork workSTations) copes with the problems arising in an extended distributed UNIX System V environment. It is a kernel-level implementation of UNIX. The kernel code of a few system calls was changed, preserving the system call interface. NEST provides location independent remote execution, process migration, and load balancing.

A prototype implementation uses a communication subsystem which is similar to the one-paths used in the DEMOS and Roscoe projects (Sects. 8.22, 8.56, resp.).

A so-called *server pool* scheme has been designed. Here, each workstation maintains a local database of servers in which it is interested in order to get CPU power as well as a database of clients to which it will offer some CPU cycles. Load distribution itself is not transparent at the user level; a user indicates that he wants to offload his programs to such a server by using a "rexec" prefix when submitting a command. Now the resulting processes are candidates for remote execution. Moreover, processes are dynamically migrated whenever the local process table is full. A dynamic reassignment scheme is planned.

CONTACT: Rakesh Agrawal, or Ahmed K. Ezzat, AT&T Bell Laboratories, Murray Hill NJ 07974, USA
e-mail: rakesh@allegra.att.com
REFERENCE: [639] – [642]

8.46   Networked Resource Discovery Project (NRDP)

The Networked Resource Discovery Project explores a number of experimental means by which users can discover the existence of resources in a large internet environment, e.g., network services, documents, retail products, current events, and people.

The designers impose three goals on their approaches to resource discovery: (a) scalability in very large environments spanning national and international networks, (b) resource space organization, (c) minimization of global administrative agreement.

CONTACT: Michael Schwartz, Dept. of Computer Science, Campus Box 430, Univ. of Colorado, Boulder CO 80309, USA
e-mail: schwartz@latour.colorado.edu
REFERENCES: [643] – [650]

8.47   NonStop

The Tandem NonStop System is a distributed computer system designed expressly for on-line transaction processing. Fault-tolerance is achieved by pairs of processes doing the same work on different processors. Processes communicate by messages.
REFERENCE: [651]

8.48   Onyx

Onyx is an object-oriented distributed programming language developed for the NEXUS distributed operating system (Sect. 5.14).

CONTACT: Anand Tripathi, Dept. of Computer Science, Univ. of Minnesota, Minneapolis MN 55455, USA
e-mail: tripathi@cs.umn.edu
REFERENCE: [652]

8.49   PHARROS

PHARROS stands for Parallel Heterogeneous Architecture for Reliable Realtime Operating Systems. It is developed at RCA's Advanced Technology Laboratories. The main goal of the project is to provide an operating system and associated distributed architecture to support applications which are distributed across next-generation networks, i.e. networks of heterogeneous parallel processors.

STATUS: In 1987 a demonstration system has been constructed. It consists of a Connection Machine, a BBN Butterfly, a VAX cluster, and a WARP, all connected by an Ethernet. A large signal processing and tracking application has been implemented on top of this system. The current status is unknown.

REFERENCE: [653]

8.50   Presto

Presto is a follow-on of the Emerald language (Sect. 6.5). Presto is an object-oriented system for parallel programming on shared-memory multiprocessors, the Sequent Symmetry. Presto is a new layer that is implemented on top of existing operating systems. It is written in C++.

CONTACT: Hank Levy, Dept. of Computer Science and Engineering, FR-35, Univ. of Washington, Seattle WA 98195, USA
e-mail: levy@cs.washington.edu

REFERENCE: [654]

8.51   R$\star$

R$\star$ is a distributed database management system. Multisite user transactions are structured in a *tree* of processes communicating over virtual circuit communication links. Beside the transaction management, distributed deadlock detection protocols are implemented. R$\star$ is running on multiple processors.

REFERENCE: [655]

8.52   Rapport

Rapport is a multimedia conferencing system developed at AT&T Bell Laboratories, Holmdel NJ 07733, USA. It runs on a collection of SUN workstations connected by an Ethernet. It makes use of the NFS file service (Sect. 2.8) which provides common names and storage for programs and data used in conferences. It is a distributed system since a collection of simultaneously active agents transparently access shared data over a network. It is planned to create a modified version of Rapport in which conferences can take place over WANs. Rapport belongs to the large group of *groupware* systems. More about groupware systems can be found in [15].

REFERENCE: [656], [657]

8.53 RNFS

RNFS is a fault-tolerant extension of the SUN NFS (Sect. 2.8) following the state machine approach. RNFS uses the ISIS broadcast protocols and toolkit routines (Sect. 6.12). The RNFS prototype provides failure transparency. This is achieved by the replication of servers as a set of identical state machines, and by the use of an atomic broadcast protocol for server communication.

RNFS shows that a distributed file system that is highly available can be run on top of a standard distributed file system (here NFS). The methodology used in this project should apply to other distributed file systems as well.

STATUS: A first prototype of RNFS was completed in 1988. This prototype supported the full NFS protocol. In the future, RNFS research will address the problem of scaling to larger networks.

CONTACT: Keith Marzullo or Frank Schmuck, Dept. of Comp. Science, Cornell Univ., Upson Hall, Ithaca NY 14853, USA

REFERENCE: [658]

8.54 Rochester's Intelligent Gateway (RIG)

Rochester's Intelligent Gateway had provided access to Arpanet services and resources as well as remote file access and other network facilities available on a local Ethernet, until its demise in 1986 because its hardware base became obsolescent. RIG is a predecessor of Accent (Sect. 4.1).

REFERENCE: [659]

8.55 ROE

ROE is a traditional distributed file system providing location and replication transparency. It uses file migration to achieve load balancing and a better disk storage allocation. To increase the availability of the files, replication is used. Through Gifford's weighted voting scheme, consistency and pessimistic concurrency control are guaranteed.

REFERENCES: [13], [660], [661]

8.56 Roscoe

Roscoe is a multi-computer distributed operating system running on a network of DEC LSI-11 machines. Roscoe was developed in 1977 and 1978 at the Computer Sciences Dept. of Wisconsin-Madison. There is a direct line between research on Roscoe and Arachne and the Charlotte system (Sects. 8.4, 4.3, resp.).

CONTACT: Raphael A. Finkel, Dept. of Computer Science, Univ. of Kentucky, Lexington KY 40506, USA

REFERENCES: [662] – [665]

8.57   RSS

RSS was developed at IBM as part of the system R project. RSS is the layer of the
system R which is responsible for transaction management and database recovery.
    CONTACT: IBM Almaden Research Center, 650 Harry Road, San Jose CA 95120-
6099, USA
    REFERENCE: [666]

8.58   RT PC Distributed Services

RT PC Distributed Services provides distributed operating system capabilities for the
AIX operating system. These include a distributed file system and distributed interpro-
cess communication. Transparency is achieved by *remote mounts* of files or file systems.
Most application programs are running without modification.
    REFERENCE: [667]

8.59   S/Net's Linda Kernel

Linda is a parallel programming language which supports shared-memory-like distributed
data structures. S/Net is a multicomputer based on a fast word-parallel bus interconnect.
A special communication kernel is implemented for this machine that supports Linda
primitives (Sect. 6.14).
    CONTACT: Nicholas Carriero, or David Gelernter, Yale Univ., resp. AT&T Bell
Laboratories, Holmdel NJ 07733, USA
    REFERENCES: [668] – [670]

8.60   Sesame

Sesame is a file system, developed at Carnegie Mellon University as part of the Spice
project. It is a successor to CMCFS (Sect. 8.12). Its main objectives are naming, authentication, authorization and data storage as well as retrieval in a network of personal
computers.
    REFERENCE: [671]

8.61   StarOS

StarOS was the first operating system developed for the Cm* multimicroprocessor computer. Cm* was designed at Carnegie Mellon University, Pittsburgh, PA. The successor
to StarOS is Medusa (Sect. 6.13).
    REFERENCE: [672]

8.62 STORK

STORK is an experimental migrating file system for computer networks developed in the Dept. of Computer Science at Purdue University, West Lafayette IN. STORK is a direct predecessor of the IBIS file system (Sect. 2.7).

CONTACT: Walter F. Tichy, Univ. of Karlsruhe, Informatik, Postfach 6980, D-7500 Karlsruhe, Germany
e-mail: tichy@ira.uka.de
REFERENCE: [673]

8.63 Thoth

Thoth was a real-time operating system which was designed to be portable over a large set of machines. Therefore, it was written in a high-level language. It provided efficient interprocess communication primitives that influenced the development of V (Sect. 4.21). The main idea was "... messages are for communication and processes are for concurrency."

CONTACT: David R. Cheriton, Distributed Systems Group, Computer Science Dept., Stanford Univ., Stanford CA 94305, USA
e-mail: cheriton@cs.stanford.edu
REFERENCE: [674], [675]

8.64 TimixV2

TimixV2 is a real-time operating system kernel developed to propose new programming language constructs for distributed real-time programming, and to demonstrate their utility by using them to program distributed robotics applications. It uses UDP/IP to allow TimixV2 applications to communicate with other applications on other processors running TimixV2 or UNIX. The TimixV2 kernel is implemented on a distributed real-time testbed consisting of DEC MicroVAXs connected by an Ethernet.

CONTACT: Insup Lee, Dept. of Computer and Information Science, Univ. of Pennsylvania, Philadelphia PA 19104-6389, USA
e-mail: lee@central.cis.upenn.edu
REFERENCES: [676] – [679]

8.65 Topaz (Taos)

Topaz is the software system for the Firefly multiprocessor (Sect. 8.31). The key facilities of Topaz are execution of existing Ultrix binaries, a remote file system (Taos), a display manager and a debug server (TTD).

Topaz provides lightweight processes and remote procedure calls for inter-address-space and inter-machine communications. The performance of the RPC is high even though relatively slow processors are used.

REFERENCE: [680]

8.66 Transparent Integrated Local and Distributed Environment (TILDE)

TILDE is a design project in the Dept. of Computer Science at Purdue University, West Lafayette IN, to build a Transparent Integrated Local and Distributed Environment. Part of the TILDE project is the IBIS file system (Sect. 2.7).
REFERENCE: [681]

8.67 Transparent Remote File System (TRFS)

The Transparent Remote File System (TRFS) provides transparent access in a distributed UNIX file system through the use of symbolic links to path names on remote machines. Thus, programs do not need to be recompiled or relinked. TRFS is a project of Integrated Solutions, Inc.
REFERENCE: [682]

8.68 Trollius (Trillium)

Trollius is an INMOS transputer (hypercube) multitasking, operating system originally implemented for the FPS T-series. It is a direct successor to *Trillium*.

Trollius combines a transputer, one or more UNIX 4.3 BSD hosts and any other UNIX-based machine into one large system. Trollius does not include implicit support for parallel processing by itself, but instead provides explicit control of all facets of machine operation including reconfiguration of the operating system. Trollius uses a simple client/server model to enable every process in the multicomputer to transparently access a UNIX-compatible filesystem. Location transparency is not provided, since filenames are prefixed, if necessary, with a node-Id. OSI-like datalink, network and transport services are provided.

The layers in Trollius are: (1) toolkit – primitive tools to work on compute nodes directly attached to host nodes, (2) kernel – multitasking, multi-priority, signal delivering, message passing kernel/runtime library – local within a node (compute or host) – does not know about other nodes, (3) network – full blown network message passing service – backbone of Trollius, (4) remote-client/server model – not unlike UNIX – handles everything from running programs on compute nodes to remote filesystem, (5) command – remote libraries wrapped up in UNIX/Trollius command programs – constitutes user interface, (6) topology – tools to configure communication links, generate route tables and boot Trollius based on physical description of multicomputer, (7) resource (not built yet) – tools to allocate nodes and map applications into machines

Trollius is a nuts-and-bolts type of system. Synchronization is on physical node identifier (integer), a priori arbitrary event (integer, match on ==), and a priori arbitrary type (integer, match on &). Abstractions and name services are built on top as desired. The only one that is built is a global associative shared memory package based on the Linda concept (Sect. 8.59). It also runs Cal Tech's CrOS grid programming library on top of Trollius.

There are **no** distributed data structures in Trollius. Spanning trees are built for multicasting but that is all controlled from one point.

The developers at the Ohio State University, Office of Research Computing, Columbus OH, and at the Cornell Theory Center, Ithaca NY, intend to operate in a mode similar to Open Software Foundation.

CONTACT: Gregory Burns or Raja Daoud, Trollius Operating System, Ohio Supercomputer Center, The Ohio State Univ., Columbus OH, USA
e-mail: gdburns@osc.edu resp. raja@ocs.edu
REFERENCE: [683]

8.69 Universal Poly-Processor with Enhanced Reliability (UPPER)

The Universal Poly-Processor with Enhanced Reliability (UPPER) is an object-oriented multicomputer system of GMD FIRST, Germany. The operating system is supported by special hardware. All communication is done by asynchronous client-server interactions. Server processes are replicated to achieve fault tolerance.

REFERENCE: [684]

8.70 Woodstock File Server (WFS)

The Woodstock File Server (WFS) is an experimental file server from Xerox PARC, Palo Alto, CA, supporting the Woodstock office system. WFS has been operational since 1975 and provides sharing of information in applications, such as an experimental telephone directory, and a library information storage and retrieval system.

REFERENCE: [685]

8.71 Xcode

The Xcode is a distributed system designed and implemented at the Politecnico di Milano, Italy, within the Cnet project. Cnet's topics under investigation include communication networks, systems programming languages, programming environments, office automation systems, and formal methods. It uses the ADA language as a basis for research. A limited ADA extension has been proposed to allow that identifiers can be bound to runtime computed modules of known interfaces.

Xcode runs on Olivetti M 40s connected by an Ethernet and provides location transparency to all private resources. It consists of an abstract machine layer *Virtual nodes* and a runtime support layer.

REFERENCE: [686]

8.72   Z-Ring File Server

The Z-Ring File Server was developed at IBM to build a testbed for the then experimental token ring network. It provides a hierarchical file system. The Z-Ring File Server was implemented in BCPL under the TRIPOS operating system.

CONTACT: IBM Zürich Research Laboratory, Säumerstr. 4, CH-8803 Rüschlikon, Switzerland

REFERENCE: [687]

Bibliography

1. Borghoff, U.M., Nast-Kolb, K.: Distributed Systems: A Comprehensive Survey. Inst. für Informatik, Techn. Univ. München, Munich, Germany, Technical Report TUM-I 8909, Nov. 1989

2. Tanenbaum, A.S., Renesse, R.van: Distributed Operating Systems. ACM Computing Surveys **17**:4, 419–470 (Dec. 1985)

3. Svobodova, L.: File Servers for Network-Based Distributed Systems. ACM Computing Surveys **16**:4, 353–398 (Dec. 1984)

4. Levy, E., Silberschatz, A.: Distributed File Systems: Concepts and Examples. ACM Computing Surveys **22**:4, 321–374 (Dec. 1990)

5. Spanier, S.: Comparing Distributed File Systems. Data Communications, pp. 173–186 (Dec. 1987)

6. Vandôme, G.: Comparative Study of some UNIX Distributed File Systems. Proc. Europ. UNIX Systems User Group Conf. Autumn '86, Manchester, UK, Sep. 1986. Buntingford Herts, UK: EUUG, pp. 73–82

7. Hatch, M.J., Katz, M., Rees, J.: AT&T's RFS and SUN's NFS – A Comparison of Heterogeneous Distributed File Systems. UNIX World, pp. 159–164 (Dec. 1985)

8. Wildgruber, R.: Verteilte UNIX-Systeme. Markt & Technik Nr. 42 (Oct. 1985)

9. Douglis, F., Kaashoek, M.F., Tanenbaum, A.S., Ousterhout, J.K.: A Comparison of Two Distributed Systems: Amoeba and Sprite. Dept. of Mathematics and Computer Science, Vrije Univ., Amsterdam, the Netherlands, Technical Report IR-230, Dec. 1990

10. Mitchell, J.G., Dion, J.: A Comparison of Two Network-Based File Servers. Communications of the ACM **25**:4, 233–245 (Apr. 1982)

11. Chin, R.S., Chanson, S.T.: Distributed Object-Based Programming Systems. ACM Computing Surveys **23**:1, 91–124 (Mar. 1991)

12. Jacqmot, C., Milgrom, E., Joosen, W., Berbers, Y.: UNIX and Load Balancing: A Survey. Proc. Europ. UNIX Systems User Group Conf. Spring '89, Brussels, Belgium, Apr. 1989. Buntingford Herts, UK: EUUG, pp. 1–15

13. Gifford, D.K.: Weighted Voting for Replicated Data. Proc. 7th ACM Symp. on Operating Systems Principles, Pacific Grove, CA, Dec. 1979. ACM SIGOPS Operating Systems Review **13**:5, pp. 150–162

14. Borghoff, U.M.: Fehlertoleranz in verteilten Dateisystemen: Eine Übersicht über den heutigen Entwicklungsstand bei den Votierungsverfahren. Informatik-Spektrum **14**:1, 15–27 (Feb. 1991)

15. Ellis, C.A., Gibbs, S.J., Rein, G.L.: Groupware – Some Issues and Experiences. Communications of the ACM **34**:1, 38–58 (Jan. 1991)

Alpine   (Sect. 2.1)

16. Brown, M.R., Kolling, K.N., Taft, E.A.: The Alpine File System. Xerox Palo Alto Research Center, Palo Alto, CA, Technical Report CSL–84–4, Oct. 1984

17. Brown, M.R., Kolling, K.N., Taft, E.A.: The Alpine File System. ACM Transactions on Computer Systems 3:4, 261–293 (Nov. 1985)

Andrew   (Sect. 2.2)

18. Accetta, M., Robertson, G., Satyanarayanan, M., Thompson, M.: The Design of a Network-Based Central File System. Dept. of Computer Science, Carnegie Mellon Univ., Pittsburgh, PA, Technical Report CMU–CS–80–134, Aug. 1980

19. Borenstein, N., Everhart, C., Rosenberg, J., Stooler, A.: A Multi-media Message System for Andrew. Proc. USENIX Conf. Winter '88, Dallas, TX, Feb. 1988. Berkeley, CA: USENIX Association

20. Dannenberg, R.B.: Resource Sharing in a Network of Personal Computer. Dept. of Computer Science, Carnegie Mellon Univ., Pittsburgh, PA, PhD thesis, Dec. 1982

21. Howard, J.H., Kazar, M.J., Menees, S.G., Nichols, D.A., Satyanarayanan, M., Sidebotham, R.N., West, M.J.: Scale and Performance in a Distributed File System. ACM Transactions on Computer Systems 6:1, 51–81 (Feb. 1988)

22. Kazar, M.L., Leverett, B.W., Anderson, O.T., Apostolides, V., Bottos, B.A., Chutani, S., Everhart, C.F., Mason, W.A., Tu, S.-T., Zayas, E.R.: DEcorum (AFS) File System Architectural Overview. Proc. USENIX Conf. Summer '90, Anaheim, CA, Jun. 1990. Berkeley, CA: USENIX Association

23. Leong, J.: Data Communication at CMU. Dept. of Computer Science, Carnegie Mellon Univ., Pittsburgh, PA, Technical Report CMU–ITC–85–043, Jul. 1985

24. Morris, J.H., Satyanarayanan, M., Conner, M.H., Howard, J.H., Rosenthal, D.S.H., Smith, F.D.: Andrew: A Distributed Personal Computing Environment. Communications of the ACM 29:3, 184–201 (Mar. 1986)

25. Nichols, D.A.: Using Idle Workstations in a Shared Computing Environment. Proc. 11th ACM Symp. on Operating Systems Principles, Austin, TX, Nov. 1987. ACM SIGOPS Operating Systems Review 21:5, pp. 5–12

26. Morris, J.H.: Make or Take Decisions in Andrew. Proc. USENIX Conf. Winter '88, Dallas, TX, Feb. 1988. Berkeley, CA: USENIX Association

27. Palay, A.J., Hansen, W.J., Kazar, M.L., Sherman, M., Wadlow, M.G., Neuendorffer, T.P., Stern, Z., Bader, M., Peters, T.: The Andrew Toolkit – An Overview. Proc. USENIX Conf. Winter '88, Dallas, TX, Feb. 1988. Berkeley, CA: USENIX Association

28. Satyanarayanan, M.: Supporting IBM PCs in a Vice/Virtue Environment. Dept. of Computer Science, Carnegie Mellon Univ., Pittsburgh, PA, Technical Report CMU-ITC-002, 1984

29. Satyanarayanan, M., Howard, J.H., Nichols, D.A., Sidebotham, R.N., Spector, A.Z., West, M.J.: The ITC Distributed File System: Principles and Design. Proc. 10th ACM Symp. on Operating Systems Principles, Orcas Island, WA, Dec. 1985. ACM SIGOPS Operating Systems Review 19:5, pp. 35–50

30. Satyanarayanan, M.: Integrating Security in a Large Distributed Environment. Dept. of Computer Science, Carnegie Mellon Univ., Pittsburgh, PA, Technical Report CMU–CS–87–179, 1987

31. Satyanarayanan, M.: Distributed File Systems. Proc. Arctic '88, An Advanced Course on Distributed Systems, Tromsø, Norway, Jul. 1988

32. Satyanarayanan, M.: On the Influence of Scale in a Distributed System. Proc. 10th IEEE Int. Conf. on Software Engineering, Singapore, Apr. 1988. Los Alamitos, CA: IEEE Comp. Soc. Press, pp. 10–18

33. Satyanarayanan, M.: Scalable, Secure, and Highly Available Distributed File Access. IEEE Computer **23**:5, 9–22 (May 1990)

34. Sidebotham, B.: Volumes – The Andrew File System Data Structure Primitive. Proc. Europ. UNIX Systems User Group Conf. Autumn '86, Manchester, UK, Sep. 1986. Buntingford Herts, UK: EUUG, pp. 473–480

35. Spector, A.Z., Kazar, M.L.: Uniting File Systems. UNIX Review (Mar. 1989)

36. DEcorum: A Technical Overview, (Feb. 1990), available from Transarc Corporation, Pittsburgh, PA

Cedar (Sect. 2.3)

37. Donahue, J.: Integration Mechanisms in Cedar. ACM SIGPLAN Notices **20**:7, 245–251 (Jul. 1985)

38. Gifford, D.K., Needham, R.M., Schroeder, M.D.: The Cedar File System. Communications of the ACM **31**:3, 288–298 (Mar. 1988)

39. Hagmann, R.: Reimplementing the Cedar File System Using Logging and Group Commit. Proc. 11th ACM Symp. on Operating Systems Principles, Austin, TX, Nov. 1987. ACM SIGOPS Operating Systems Review **21**:5, pp. 155–162

40. Schmidt, E.E.: Controlling Large Software Development in a Distributed Environment. Xerox Palo Alto Research Center, Palo Alto, CA, Technical Report CLS–82–7, Dec. 1982

41. Schroeder, M.D., Gifford, D.K., Needham, R.M.: A Cashing File System for a Programmer's Workstation. Xerox Palo Alto Research Center, Palo Alto, CA, Technical Report CLS–85–7, Nov. 1985

42. Swinehart, D.C., Zellweger, P.T., Hagmann, R.B.: The Structure of Cedar. ACM SIGPLAN Notices **20**:7, 230–244 (Jul. 1985)

43. Swinehart, D.C., et al.: A Structural View of the Cedar Programming Environment. ACM Transactions on Programming Languages and Systems **8**:4, 419–490 (Oct. 1986)

44. Teitelman, W.: The Cedar Programming Environment: A Midterm Report and Examination. Xerox Palo Alto Research Center, Palo Alto, CA, Technical Report CLS–83–11, Jul. 1984

45. Thacher, C., et al.: Alto: A Personal Computer. Xerox Palo Alto Research Center, Palo Alto, CA, Technical Report CLS–79–11, Aug. 1979

Coda (Sect. 2.4)

46. Satyanarayanan, M., Kistler, J.J., Kumar, P., Okasaki, M.E., Siegel, E.H., Steere, D.C.: Coda: A highly available File System for a Distributed Workstation Environment. IEEE Transactions on Computers c–**39**:4, 447–459 (Apr. 1990)

47. Satyanarayanan, M., Siegel, E.H.: Parallel Communication in a Large Distributed Environment. IEEE Transactions on Computers c–**39**:3, 328–348 (Mar. 1990)

48. Steere, D.C., Kistler, J.J., Satyanarayanan, M.: Efficient User-Level File Cache Management on the Sun Vnode Interface. Proc. USENIX Conf. Summer '90, Anaheim, CA,

Jun. 1990. Berkeley, CA: USENIX Association, also available as Carnegie Mellon Univ., Pittsburgh, PA, Technical Report CMU-CS-90-126, April, 1990

EFS   (Sect. 2.5)

49. Cole, C.T., Flinn, P.B., Atlas, A.B.: An Implementation of an Extended File System for UNIX. Proc. USENIX Conf. Summer '85, Portland, OR, Jun. 1985. Berkeley, CA: USENIX Association, pp. 131–149

HARKYS   (Sect. 2.6)

50. Baldi, A., Stefanelli, L.: HARKYS – A New Network File System Approach. Proc. Europ. UNIX Systems User Group Conf. Autumn '86, Manchester, UK, Sep. 1986. Buntingford Herts, UK: EUUG, pp. 163–173

IBIS   (Sect. 2.7)

51. Ruan, Z., Tichy, W.F.: Performance Analysis of File Replication Schemes in Distributed Systems. Proc. ACM SIGMETRICS Conf. on Measurement and Modeling of Computer Systems, 1987. ACM SIGMETRICS Performance Evaluation Review, **15**, pp. 205–215

52. Tichy, W.F.: Towards a Distributed File System. Proc. USENIX Conf. Summer '84, Salt Lake City, UT, Jun. 1984. Berkeley, CA: USENIX Association, pp. 87–97

NFS   (Sect. 2.8)

53. Fraser-Campbell, B., Rosen, M.B.: An Implementation of NFS under System V.2. Proc. Europ. UNIX Systems User Group Conf. Spring '86, Florence, Italy, Apr. 1986. Buntingford Herts, UK: EUUG

54. Lyon, B.: Sun Remote Procedure Call Specification. Sun Microsystems, Inc., Mountain View, CA, Technical Report, 1984

55. Lyon, B.: Sun External Data Representation Specification. Sun Microsystems, Inc., Mountain View, CA, Technical Report, 1984

56. Lyon, B., Sager, G., Chang, J.M., Goldberg, D., Kleiman, S., Lyon, T., Sandberg, R., Walsh, D., Weiss, P.: Overview of the SUN Network File System. Proc. USENIX Conf. Winter '85, Dallas, TX, Jan. 1985. Berkeley, CA: USENIX Association, pp. 1–8

57. Rosen, M.B., Wilde, M.J., Fraser-Campbell, B.: NFS Portability. Proc. USENIX Conf. Summer '86, Atlanta, GA, Jun. 1986. Berkeley, CA: USENIX Association

58. Sandberg, R.: The SUN Network File System: Design, Implementation and Experience. SUN Microsystems, Inc., Mountain View, CA, Technical Report, 1985

59. Sandberg, R.: Sun Network Filesystem Protocol Specification. Sun Microsystems, Inc., Mountain View, CA, Technical Report, 1985

60. Sandberg, R., Goldberg, D., Kleiman, S., Walsh, D., Lyone, B.: Design and Implementation of the SUN Network File System. Proc. USENIX Conf. Summer '85, Portland, OR, Jun. 1985. Berkeley, CA: USENIX Association, pp. 119–130

61. Remote Procedure Call Reference Manual. SUN Microsystems, Inc., Mountain View, CA, Oct. 1984

62. External Data Representation Reference Manual. SUN Microsystems, Inc., Mountain View, CA, Oct. 1984

63. Remote Procedure Call Protocol Specification. SUN Microsystems, Inc., Mountain View, CA, Oct. 1984

64. Network Programming. SUN Microsystems, Inc., Mountain View, CA, Part Number: 800-1779-10, Rev. A, May 1988

65. NFS: Network File System Protocol Specification. SUN Microsystems, Inc., Mountain View, CA, RFC 1094, Network Information Center, SRI Int., Mar. 1989

66. Weiss, P.: Yellow Pages Protocol Specification. Sun Microsystems, Inc., Mountain View, CA, Technical Report, 1985

67. West, A.: The SUN Network File System, NFS – Business Overview. SUN Microsystems, Inc., Mountain View, CA, Technical Report, 1985

RFS   (Sect. 2.9)

68. Bach, M.J., Luppi, M.W., Melamed, A.S., Yueh, K.: A Remote File Cache for RFS. Proc. USENIX Conf. Summer '87, Phoenix, AZ, Jun. 1987. Berkeley, CA: USENIX Association, pp. 273–280

69. Charlock, H.: RFS in SunOS. Proc. USENIX Conf. Summer '87, Phoenix, AZ, Jun. 1987. Berkeley, CA: USENIX Association

70. Hamilton, R., Krzyzanowski, P., Padovano, M., Purdome, M.: An Administrator's View of RFS. AT&T Information Systems (Jun. 1986)

71. Kingston, D.P.: Remote File Systems on UNIX. Proc. Europ. UNIX Systems User Group Conf. Autumn '85, Copenhagen, Denmark, Sep. 1985. Buntingford Herts, UK: EUUG, pp. 77–93

72. Rifkin, A.P., Forbes, M.P., Hamilton, R.L., Sabrio, M., Shah, S., Yueh, K.: RFS Architectural Overview. EUUGN 6:2, 13–23 (1986)

73. Ritchie, D.M.: A Stream Input-Output System. AT&T Bell Techn. Journal 63:8 (Oct. 1984)

S-/F-UNIX   (Sect. 2.10)

74. Chesson, G.L.: Datakit Software Architecture. Proc. ICC, Boston, MA, Jun. 1979, pp. 20.2.1–20.2.5

75. Fraser, A.G.: Datakit – A Modular Network for Synchronous and Asynchronous Traffic. Proc ICC, Boston, MA, Jun. 1979, pp. 20.1.1–20.1.3

76. Luderer, G.W.R., Che, H., Haggerty, J.P., Kirslis, P.A., Marshall, W.T.: A Distributed UNIX System Based on a Virtual Circuit Switch. Proc. 8th ACM Symp. on Operating Systems Principles, Asilomar, CA, Dec. 1981. ACM SIGOPS Operating Systems Review 15:5, pp. 160–168

77. Luderer, G.W.R., Che, H., Marshall, W.T.: A Virtual Circuit Switch as the Basis for a Distributed System. Proc. 7th ACM/IEEE Symp. on Data Communications, Oct. 1981, pp. 27–29

Spritely-NFS   (Sect. 2.11)

78. Srinivasan, V., Mogul, J.C.: Spritely NFS: Experiments with Cache-Consistency Protocols. Proc. 12th ACM Symp. on Operating Systems Principles, The Wigwam Litchfield Park, AZ, Dec. 1989. ACM SIGOPS Operating Systems Review 23:5, pp. 45–57

79. Srinivasan, V., Mogul, J.C.: Spritely NFS: Experiments with and Implementation of Cache-Consistency Protocols. DEC Western Research Lab., Technical Report 89/5, Mar. 1989

VAXcluster (Sect. 2.12)

80. Kronenberg, N.P., Levy, H.M., Strecker, W.D.: VAXclusters: A Closely–Coupled Distributed System. ACM Transactions on Computer Systems 4:2, 130–146 (May 1986)

XDFS (Sect. 2.13)

81. Sturgis, H.E., Mitchell, J.G., Israel, J.: Issues in the Design and Use of a Distributed File System. ACM SIGOPS Operating Systems Review 14:3, 55–69 (Jul. 1980)

DOMAIN (Sect. 3.1)

82. Lazowska, E.D., Zahorjan, J., Cheriton, D.R., Zwaenepooel, W.: File Access Performance of Diskless Workstations. Dept. of Computer Science, Univ. of Washington, Seattle, WA, Technical Report 84–06–01, Jun. 1984
83. Leach, P.J., Stumpf, B.L., Hamilton, J.A., Levine, P.H.: UID's as Internal Names in a Distributed File System. Proc. 1st ACM Symp. on Principles of Distributed Computing, Ottawa, Canada, 1982, pp. 34–41
84. Leach, P.J., Levine, P.H., Douros, B.P., Hamilton, J.A., Nelson, D.L., Stumpf, B.L.: The Architecture of an Integrated Local Network. IEEE Journal on Selected Areas in Communications SAC–1:5, 842–857 (Nov. 1983)
85. Leach, P.J., Levine, P.H., Hamilton, J.A., Stumpf, B.L.: The File System of an Integrated Local Network. Proc. ACM Computer Science Conf., New Orleans, LA, Mar. 1985, pp. 309–324
86. Leblang, D.B., Chase, R.P.: Computer-Aided Software Engineering in a Distributed Workstation Environment. In: Henderson, P. (ed.): Proc. ACM SIGSOFT/SIGPLAN Software Engineering Symp. on Practical Software Development Environments, Pittsburgh, PA, Apr. 1984. ACM SIGPLAN Notices 19:5, pp. 104–112
87. Levine, P.H.: The Apollo DOMAIN Distributed File System. In: Paker, Y., Banatre, J.-P., Bozyigit, M. (eds.): Distributed Operating Systems: Theory and Practice, NATO ASI series Vol. F28, 1987. Berlin, Heidelberg, New York: Springer-Verlag, pp. 241–260
88. Levine, P.H.: The DOMAIN System. Proc. 2nd ACM SIGOPS Europ. Workshop on Making Distributed Systems Work, Amsterdam, the Netherlands, Sep. 1986. ACM SIGOPS Operating Systems Review 21:1, 49–84, Jan. 1987
89. Nelson, D.L., Leach, P.J.: The Evolution of the Apollo DOMAIN. Proc. IEEE COMPCON Conf. Spring '84, San Francisco, CA, Feb. 1984. Los Alamitos, CA: IEEE Comp. Soc. Press, pp. 132–141, also in Proc. HICSS, pp. 470–479, 1984
90. Nelson, D.L., Leach, P.J.: The Architecture and Applications of the Apollo DOMAIN. IEEE Computer Graphics and Applications (Apr. 1984)
91. Rees, J., Levine, P.H., Mischkin, N., Leach, P.J.: An Extensible I/O System. Proc. USENIX Conf. Summer '86, Atlanta, GA, Jun. 1986. Berkeley, CA: USENIX Association, pp. 114–125
92. Rees, J., Olson, M., Sasidhar, J.: A Dynamically Extensible Streams Implementation. Proc. USENIX Conf. Summer '87, Phoenix, AZ, Jun. 1987. Berkeley, CA: USENIX Association, pp. 199–207

Helix   (Sect. 3.2)

93. Fridrich, M., Older, W.: Helix: The Architecture of a Distributed File System. Proc. 4th IEEE Int. Conf. on Distributed Computing Systems, San Francisco, CA, May 1984. Los Alamitos, CA: IEEE Comp. Soc. Press, pp. 422–431

94. Fridrich, M., Older, W.: Helix: The Architecture of the XMS Distributed File System. IEEE Software **2**:3, 21–29 (May 1985)

95. Gammage, N., Casey, L.: XMS: A Rendezvous-Based Distributed System Software Architecture. IEEE Software **2**:3, 9–19 (May 1985)

SWALLOW   (Sect. 3.3)

96. Arens, G.C.: Recovery of the SWALLOW Repository. MIT Lab. for Computer Science, Cambridge, MA, Technical Report MIT/LCS/TR–252, Jan. 1981

97. Reed, D.P., Svobodova, L.: SWALLOW: A Distributed Data Storage System for a Local Network. In: West, A., Janson, P. (eds.): Proc. on Local Networks for Computer Communications, 1981. Amsterdam, New York: North-Holland, pp. 355–373

98. Svobodova, L.: Management of the Object Histories in the SWALLOW Repository. MIT Lab. for Computer Science, Cambridge, MA, Technical Report MIT/LCS/TR–243, Jul. 1980

Accent   (Sect. 4.1)

99. Fitzgerald, R., Rashid, R.F.: The Integration of Virtual Memory Management and Interprocess Communication in Accent. ACM Transactions on Computer Systems **4**:2, 147–177 (May 1986)

100. Rashid, R.F.: An Inter-Process Communication Facility for UNIX. Dept. of Computer Science, Carnegie Mellon Univ., Pittsburgh, PA, Technical Report CMU–CS–80–124, Feb. 1980

101. Rashid, R.F., Robertson, G.G.: Accent: A Communication Oriented Network Operating System Kernel. Proc. 8th ACM Symp. on Operating Systems Principles, Asilomar, CA, Dec. 1981. ACM SIGOPS Operating Systems Review **15**:5, pp. 64–75

102. Rashid, R.F.: The Accent Kernel Interface Manual. Dept. of Computer Science, Carnegie Mellon Univ., Pittsburgh, PA, Technical Report, Jan. 1983

103. Rashid, R.F.: Experiences with the Accent Network Operating System. In: Müller, G., Blanc, R.P. (eds.): Proc. Int. Seminar on Networking in Open Systems, Oberlech, Austria, Aug. 1986. Lecture Notes in Computer Science **248**. Berlin, Heidelberg, New York: Springer-Verlag, pp. 270–295

104. Zayas, E.: Attacking the Process Migration Bottleneck. Proc. 11th ACM Symp. on Operating Systems Principles, Austin, TX, Nov. 1987. ACM SIGOPS Operating Systems Review **21**:5, pp. 13–24

CDCS   (Sect. 4.2)

105. Bacon, J.M., Leslie, I.M., Needham, R.M.: Distributed Computing with a Processor Bank. In: Schröder-Preikschat, W., Zimmer, W. (eds.): Proc. Europ. Workshop on Progress in Distributed Operating Systems and Distributed Systems Management, Berlin, Germany, Apr. 1989. Lecture Notes in Computer Science **433**. Berlin, Heidelberg, New York: Springer-Verlag, pp. 147–161

106. Dion, J.: The Cambridge File Server. ACM SIGOPS Operating Systems Review **14**:4, 26–35 (Oct. 1980)

107. Dion, J.: Reliable Storage in a Local Network. Computer Lab., Univ. of Cambridge, UK, Technical Report 16, 1981

108. Garnett, N.H., Needham, R.M.: An Asynchronous Garbage Collector for the Cambridge File Server. ACM SIGOPS Operating Systems Review **14**:4, 36–40 (Oct. 1980)

109. Herbert, A.J., Needham, R.M.: The User Interface to the Cambridge Distributed System. Proc. 8th ACM Symp. on Operating Systems Principles, Asilomar, CA, Dec. 1981. ACM SIGOPS Operating Systems Review **15**:5

110. Hopper, A.: The Cambridge Ring – A Local Network. In: Hanna, F.K. (ed.): Advanced Techniques for Microprocessor Systems, 1980. Stevenage, UK, New York: P. Peregrinus ltd., pp. 67–71

111. Needham, R.M., Herbert, A.J.: The Cambridge Distributed System. Reading, MA: Addison-Wesley, 1982

112. Wilkes, M.V., Needham, R.M.: The Cambridge Model Distributed System. ACM SIGOPS Operating Systems Review **14**:1, 21–29 (Jan. 1980)

Charlotte (Sect. 4.3)

113. Artsy, Y., Chang, H., Finkel, R.A.: Interprocess Communication in Charlotte. IEEE Software **4**:1, 22–28 (Jan. 1987)

114. Artsy, Y., Chang, H.-Y., Finkel, R.: Designing a Process Migration Facility – The Charlotte Experience. IEEE Computer **22**:9, 47–56 (Sep. 1989)

115. DeWitt, D.J., Finkel, R.A., Solomon, M.H.: The Crystal Multicomputer: Design and Implementation Experience. IEEE Transactions on Software Engineering **SE–13**:8, 953–966 (Aug. 1987)

116. Finkel, R.A., Solomon, M.H., DeWitt, D.J., Landweber, L.: The Charlotte Distributed Operating System. Dept. of Computer Science, Univ. of Wisconsin-Madison, Madison, WI, Technical Report 502, Oct. 1983

117. Finkel, R.A., Scott, M.L., Artsy, Y., Chang, H.: Experience with Charlotte: Simplicity and Function in a Distributed Operating System. IEEE Transactions on Software Engineering **SE–15**:6, 676–685 (Jun. 1989)

DEMOS/MP (Sect. 4.4)

118. Miller, B.P., Presotto, D.L., Powell, M.L.: DEMOS/MP: The Development of a Distributed System. Software – Practice and Experience **17**:4, 277–290 (Apr. 1987)

119. Powell, M.L.: The DEMOS File System. Proc. 6th ACM Symp. on Operating Systems Principles, Purdue Univ., West Lafayette, IN, Nov. 1977. ACM SIGOPS Operating Systems Review **11**:5, pp. 33–42

120. Powell, M.L., Miller, B.P.: Process Migration in DEMOS/MP. Proc. 9th ACM Symp. on Operating Systems Principles, Bretton Woods, NH, Oct. 1983. ACM SIGOPS Operating Systems Review **17**:5, pp. 110–119

121. Powell, M.L., Presotto, D.L.: PUBLISHING: A Reliable Broadcast Communication Mechanism. Proc. 9th ACM Symp. on Operating Systems Principles, Bretton Woods, NH, Oct. 1983. ACM SIGOPS Operating Systems Review **17**:5, pp. 100–109

DIOS   (Sect. 4.5)

122. Marquez, J.A., Cunha, J.P., Cunha, A., Verissimo, P.: SMD – A Modular Architecture for a Distributed System. Proc. SEIR-2, Univ. de Santiago de Compostela, Portugal, Sep. 1982, pp. 310–323

123. Marquez, J.A., Cunha, J.P., Guedes, P., Guimaraes, N., Cunha, A.: The Distributed Operating System of the SMD Project. Software – Practice and Experience 18:9, 859–877 (Sep. 1988)

DACNOS   (Sect. 4.6)

124. Drachenfels, H.v.: Ein Orientierungsdienst im Informationsverarbeitungsverbund. Proc. 9th NTG/GI Conf. Architektur und Betrieb von Rechensystemen, Stuttgart, Germany, Mar. 1986. NTG-Fachberichte **92**. Berlin, Offenbach: VDE-Verlag, pp. 321–331

125. Eberle, H., Schmutz, H.: NOS Kernels for Heterogeneous Environments. In: Müller, G., Blanc, R.P. (eds.): Proc. Int. Seminar on Networking in Open Systems, Oberlech, Austria, Aug. 1986. Lecture Notes in Computer Science **248**. Berlin, Heidelberg, New York: Springer-Verlag, pp. 270–295

126. Eberle, H., Geihs, K., Schill, A., Schmutz, H., Schöner, B.: Generic Support for Distributed Processing in Heterogeneous Networks. In: Krüger, G., Müller, G. (eds.): HECTOR, Vol. II: Basic Projects, 1988. Berlin, Heidelberg, New York: Springer-Verlag, pp. 80–109

127. Förster, C.: Task Setup Service for Distributed Systems. In: Gerner, N., Spaniol, O. (eds.): Proc. 5th GI/NTG Conf. – Communication in Distributed Systems, Aachen, Germany, Feb. 1987. Informatik-Fachberichte **130**. Berlin, Heidelberg, New York: Springer-Verlag, pp. 154–166

128. Förster, C.: Controlling Distributed User Tasks in Heterogeneous Networks. In: Krüger, G., Müller, G. (eds.): HECTOR, Vol. II: Basic Projects, 1988. Berlin, Heidelberg, New York: Springer-Verlag, pp. 183–199

129. Geihs, K., Seifert, M.: Validation of a Protocol for Application Layer Services. In: Krüger, G., Müller, G. (eds.): HECTOR, Vol. II: Basic Projects, 1988. Berlin, Heidelberg, New York: Springer-Verlag, pp. 306–320

130. Geihs, K., Seifert, M.: Automated Validation of a Co-operation Protocol for Distributed Systems. Proc. 6th IEEE Int. Conf. on Distributed Computing Systems, Cambridge, MA, May 1986. Los Alamitos, CA: IEEE Comp. Soc. Press, pp. 436–443

131. Geihs, K., Staroste, R., Eberle, H.: Operating System Support for Heterogeneous Distributed Systems. In: Gerner, N., Spaniol, O. (eds.): Proc. 5th GI/NTG Conf. – Communication in Distributed Systems, Aachen, Germany, Feb. 1987. Informatik-Fachberichte **130**. Berlin, Heidelberg, New York: Springer-Verlag, pp. 178–189

132. Hollberg, U., Krämer, E.: Transparent Access to Remote Files in Heterogeneous Networks. In: Krüger, G., Müller, G. (eds.): HECTOR, Vol. II: Basic Projects, 1988. Berlin, Heidelberg, New York: Springer-Verlag, pp. 140–168

133. Mattes, B., Drachenfels, H.v.: Directories and Orientation in Heterogeneous Networks. In: Krüger, G., Müller, G. (eds.): HECTOR, Vol. II: Basic Projects, 1988. Berlin, Heidelberg, New York: Springer-Verlag, pp. 110–125

134. Mattes, B.: Authentication and Authorization in Resource Sharing Networks. In: Krüger, G., Müller, G. (eds.): HECTOR, Vol. II: Basic Projects, 1988. Berlin, Heidelberg, New York: Springer-Verlag, pp. 126–139

135. Öchsle, R.: A Remote Execution Service in a Heterogeneous Network. In: Krüger, G., Müller, G. (eds.): HECTOR, Vol. II: Basic Projects, 1988. Berlin, Heidelberg, New York: Springer-Verlag, pp. 169–182

136. Seifert, M., Eberle, H.: Remote Service Call: A Network Operating System Kernel for Heterogeneous Distributed Systems. Proc. 9th NTG/GI Conf. Architektur und Betrieb von Rechensystemen, Stuttgart, Germany, Mar. 1986. NTG-Fachberichte **92**. Berlin, Offenbach: VDE-Verlag, pp. 292–305

137. Seifert, M., Eberle, H.: Remote Service Call: A NOS Kernel and its Protocols. Proc. 8th ICCC, Sep. 1986, pp. 675–680

138. Staroste, R., Schmutz, H., Wasmund, M., Schill, A., Stoll, W.: A Portability Environment for Communication Software. In: Krüger, G., Müller, G. (eds.): HECTOR, Vol. II: Basic Projects, 1988. Berlin, Heidelberg, New York: Springer-Verlag, pp. 51–79

139. Wettstein, H., Schmutz, H., Drobnik, O.: Cooperative Processing in Heterogeneous Computer Networks. In: Krüger, G., Müller, G. (eds.): HECTOR, Vol. II: Basic Projects, 1988. Berlin, Heidelberg, New York: Springer-Verlag, pp. 32–50

DUNE (Sect. 4.7)

140. Alberi, J.L., Pucci, M.F.: The DUNE Distributed Operating System. Proc. 1988 Using National Conf., Denver, CO, 1988, also available as Bellcore Technical Report TM-ARH-010641, 1987

141. Pucci, M.F., Alberi, J.L.: Optimized Communication in an Extended Remote Procedure Call Model. Computer Architecture News, pp. 37–44 (Sep. 1988), also available as Bellcore Technical Report, 1987

142. Pucci, M.F., Alberi, J.L.: Using Hints in DUNE Remote Procedure Calls. Computing Systems **3**:1, 48–68 (Winter 1990)

DUNIX (Sect. 4.8)

143. Frieder, O., Litman, A., Segal, M.E.: DUNIX: Distributed Operating Systems Education via Experimentation. Proc. Euromicro, London, UK, 1989

144. Litman, A.: DUNIX – A Distributed UNIX System. Proc. Europ. UNIX Systems User Group Conf. Autumn '86, Manchester, UK, Sep. 1986. Buntingford Herts, UK: EUUG, pp. 23–31, ACM SIGOPS Operating Systems Review **22**:1, 42–50, Jan. 1988

Freedomnet (Sect. 4.9)

145. Mitchell, M., Moat, K., Truscott, T., Warren, R.: Invoking System Calls from within the UNIX Kernel. Proc. USENIX Conf. Winter '88, Dallas, TX, Feb. 1988. Berkeley, CA: USENIX Association

146. Truscott, T., Warren, R.B., Moat, K.: A State-wide Distributed Computing System. Proc. USENIX Conf. Summer '86, Atlanta, GA, Jun. 1986. Berkeley, CA: USENIX Association

147. Warren, R.B., Truscott, T.R., Moat, K., Mitchell, M.: Distributed Computing using RTI's FREEDOMNET in a Heterogeneous UNIX Environment. Proc. UNIFORUM '87 Conf., Washington, DC, 1987

HERMIX (Sect. 4.10)

148. Berbers, Y., Busse, P., Decker, B. De, Deurwaerder, A. De, Huens, J., Moons, H., Verbaeten, P.: The HERMIX Manifesto. Katholieke Univ. Leuven, Belgium, Technical Report CW-35, 1984

149. Berbers, Y., Verbaeten, P.: The Structure of the HERMIX Distributed System. Proc. 2nd ACM SIGOPS Europ. Workshop on Making Distributed Systems Work, Amsterdam, the Netherlands, Sep. 1986. ACM SIGOPS Operating Systems Review 21:1, 49–84, Jan. 1987

150. Berbers, Y.: Design of the HERMIX Distributed Operating System. Dept. Computerwetenschappen, K.U. Leuven, Belgium, PhD thesis, Dec. 1987

151. Berbers, Y., Decker, B. De, Moons, H., Verbaeten, P.: The Design of the HERMIX Distributed System. Proc. 34th ISMM Int. Symp. Mini and Microcomputers, Lugano, Switzerland, 1987

152. Berbers, Y., Verbaeten, P.: Design of the HERMIX Operating System: Structural Aspects. Microprocessing and Microprogramming 24, 187–194 (1988)

153. Berbers, Y., Verbaeten, P.: A Hierarchical Description of the HERMIX Distributed Operating System. Proc. ACM Symp. on Personal and Small Computers, Cannes, France, 1988, pp. 98–106

154. Berbers, Y., Joosen, W., Verbaeten, P.: On the Use of Load Balancing Mechanisms for Fault Tolerance Support. Proc. 4th ACM SIGOPS Europ. Workshop on Fault Tolerance Support in Distributed Systems, Bologna, Italy, 1990

155. Berbers, Y., Verbaeten, P.: Servers, Processes and Subprocesses: A Critical Evaluation. Proc. 5th Conf. on Information Technology, Jerusalem, Israel, 1990

156. Decker, B. De: Communication in Distributed Systems: The HERMIX Model. Katholieke Univ. Leuven, Belgium, PhD thesis, 1988

157. Moons, H., Verbaeten, P.: Software Development in Server-Oriented Systems – The HERMIX Approach. Microprocessing and Microprogramming 24, 179–186 (1988)

JASMIN (Sect. 4.11)

158. Fishman, D.H., Lai, M.-Y., Wilkinson, K.: An Overview of the JASMIN Database Machine. In: Yormark, B. (ed.): Proc. ACM SIGMOD '84 Annual Conf., Boston, MA, Jun. 1984. ACM SIGMOD Record 14:2, pp. 234–239

159. Lai, M.-Y., Wilkinson, K., Lanin, V.: On Distributing JASMIN's Optimistic Multiversioning Page Manager. IEEE Transactions on Software Engineering SE–15:6, 696–704 (Jun. 1989)

160. Lee, H., Premkumar, U.: The Architecture and Implementation of the Distributed JASMIN Kernel. Bellcore, Morristown, NJ, Technical Report TM–ARH–000324, Oct. 1984

161. Uppaluru, P., Wilkinson, W.K., Lee, H.: Reliable Servers in the JASMIN Distributed System. Proc. 7th IEEE Int. Conf. on Distributed Computing Systems, Berlin, Germany, Sep. 1987. Los Alamitos, CA: IEEE Comp. Soc. Press, pp. 105–111

LOCUS (Sect. 4.12)

162. Goldberg, A., Popek, G.J.: Measurement of the Distributed Operating System LOCUS. Univ. of California, Los Angeles, CA, Technical Report, 1983

163. Moore, J.D.: Simple Nested Transactions in LOCUS. Dept. of Computer Science, Univ. of California, Los Angeles, CA, Master's thesis, 1982

164. Mueller, E., Moore, J., Popek, G.J.: A Nested Transaction System for LOCUS. Proc. 9th ACM Symp. on Operating Systems Principles, Bretton Woods, NH, Oct. 1983. ACM SIGOPS Operating Systems Review **17**:5, pp. 71–89

165. Parker, D.S., Popek, G.J., Rudisin, G., Stoughton, A., Walker, B., Walton, E., Chow, J., Edwards, D., Kiser, S., Kline, C.: Detection of Mutual Inconsistency in Distributed Systems. IEEE Transactions on Software Engineering SE–9:2, 240–247 (May 1983)

166. Popek, G.J., Walker, B., Chow, J., Edwards, D., Kline, C., Rudisin, G., Thiel, G.: LOCUS: A Network Transparent, High Reliability Distributed System. Proc. 8th ACM Symp. on Operating Systems Principles, Asilomar, CA, Dec. 1981. ACM SIGOPS Operating Systems Review **15**:5, pp. 169–177

167. Popek, G.J., Walker, B.J. (eds.): The LOCUS Distributed System Architecture. Cambridge, MA, London, UK: MIT Press, 1985

168. Sheltzer, A.B., Lindell, R., Popek, G.J.: Name Service Locality and Cache Design in a Distributed Operating System. Proc. 6th IEEE Int. Conf. on Distributed Computing Systems, Cambridge, MA, May 1986. Los Alamitos, CA: IEEE Comp. Soc. Press, pp. 515–523

169. Sheltzer, A.B., Popek, G.J.: Internet LOCUS: Extending Transparency to an Internet Environment. IEEE Transactions on Software Engineering SE–12:11, 1067–1075 (Nov. 1986)

170. Walker, B., Popek, G.J., English, R., Kline, C., Thiel, G.: The LOCUS Distributed Operating System. Proc. 9th ACM Symp. on Operating Systems Principles, Bretton Woods, NH, Oct. 1983. ACM SIGOPS Operating Systems Review **17**:5, pp. 49–70

171. Walker, B.: Issues of Network Transparency and File Replication in the Distributed Filesystem Component of LOCUS. Dept. of Computer Science, Univ. of California, Los Angeles, CA, PhD thesis, 1983

172. Walker, B., Popek, G.J., English, R., Kline, C., Thiel, G.: The LOCUS Distributed Operating System. ACM SIGOPS Operating Systems Review **17**:5, 49–70 (Oct. 1983)

MACH (Sect. 4.13)

173. Acetta, M., Baron, R., Bolosky, W., Golub, D., Rashid, R., Tevanian, A., Young, M.: Mach: A New Kernel Foundation for UNIX Development. Proc. USENIX Conf. Summer '86, Atlanta, GA, Jun. 1986. Berkeley, CA: USENIX Association, pp. 93–113

174. Baron, R.V., Rashid, R.F., Siegel, E.H, Tevanian, A., Young, M.W.: Mach-1: A Multiprocessor Oriented Operating System and Environment. New Computing Environments: Parallel, Vector and Systolic, 1986. SIAM, pp. 80–99

175. Baron, R.V., Black, D., Bolosky, W., Chew, J., Golub, D.B., Rashid, R.F., Tevanian, A., Young, M.W.: Mach Kernel Interface Manual. Dept. of Computer Science, Carnegie Mellon Univ., Pittsburgh, PA, Technical Report, Oct. 1987

176. Black, D.L.: Scheduling Support for Concurrency and Parallelism in the Mach Operating System. IEEE Computer **23**:5, 35–43 (May 1990)

177. Draves, R., Jones, M., Thompson, M.: MIG – The Mach Interface Generator. Dept. of Computer Science, Carnegie Mellon Univ., Pittsburgh, PA, Technical Report, Feb. 1988

178. Jones, M.B., Rashid, R.F., Thompson, M.R.: Matchmaker: An Interface Specification Language for Distributed Processing. Proc. 12th ACM SIGACT/SIGPLAN Annual Symp. on Principles of Programming Languages, 1985. New York: ACM

179. Jones, M.B., Rashid, R.F.: Mach and Matchmaker: Kernel and Language Support for Object-Oriented Distributed Systems. Proc. 1st ACM SIGPLAN OOPSLA '86 Conf. on Object-Oriented Programming Systems, Languages and Applications, Portland, OR, Sep. 1986. ACM SIGPLAN Notices 21:11, pp. 67–77

180. Rashid, R.F.: From RIG to Accent to Mach: The Evolution of a Network Operating System. Proc. 1st ACM/IEEE AFIPS Fall Joint Computer Conf., Dallas, TX, Nov. 1986, pp. 1128–1137

181. Rashid, R.: Mach: Layered Protocols vs. Distributed Systems. Proc. IEEE COMPCON Conf. Spring '87, San Francisco, CA, Feb. 1987. Los Alamitos, CA: IEEE Comp. Soc. Press, pp. 82–85

182. Rashid, R.F., Tevanian, A., Young, M.W., Golub, D.B., Baron, R., Black, D., Bolosky, W., Chew, J.: Machine Independent Virtual Memory Management for Paged Uniprocessors and Multiprocessor Architectures. ACM SIGOPS Operating Systems Review 21:4, 31–39 (Oct. 1987)

183. Sansom, R.D., Julin, D.P., Rashid, R.F.: Extending a Capability Based System into a Network Environment. Proc. ACM SIGCOMM Symp. in Communications Architectures and Protocols, Stowe, VT, Aug. 1986. ACM SIGCOMM Computer Communication Review 16:3, pp. 265–274

184. Tevanian, A., Rashid, R.F., Young, M.W., Golub, D.B., Thompson, M.R., Bolosky, W., Sanzi, R.: A UNIX Interface for Shared Memory and Memory Mapped Files under Mach. Proc. USENIX Conf. Summer '87, Phoenix, AZ, Jun. 1987. Berkeley, CA: USENIX Association, pp. 53–67

185. Tevanian, A., Rashid, R.F., Young, M.W., Golub, D.B., Black, D.L., Cooper, E.C.: Mach Threads and the UNIX Kernel: The Battle for Control. Proc. USENIX Conf. Summer '87, Phoenix, AZ, Jun. 1987. Berkeley, CA: USENIX Association, pp. 185–197

186. Tevanian, A., Rashid, R.F.: Mach: A Basis for Future UNIX Development. Dept. of Computer Science, Carnegie Mellon Univ., Pittsburgh, PA, Pittsburgh, PA 15213, Technical Report, Jun. 1987

187. Young, M., Tevanian, A., Rashid, R., Golub, D., Eppinger, J., Chew, J., Bolosky, W., Black, D., Baron, R.: The Duality of Memory and Communication in the Implementation of a Multiprocessor Operating System. Proc. 11th ACM Symp. on Operating Systems Principles, Austin, TX, Nov. 1987. ACM SIGOPS Operating Systems Review 21:5, pp. 63–76

MOS (Sect. 4.14)

188. Barak, A., Drezner, Z.: Distributed Algorithms for the Average Load of a Multicomputer. Computing Research Lab., Univ. of Michigan, Technical Report CRL–TR–17–84, Mar. 1984

189. Barak, A., Shiloh, A.: A Distributed Load-Balancing Policy for a Multicomputer. Software – Practice and Experience 15:9, 901–913 (Sep. 1985)

190. Barak, A., Litman, A.: A Multicomputer Distributed Operating System. Software – Practice and Experience 15:8, 727–737 (Aug. 1985)

191. Barak, A., Paradise, O.G.: MOS – A Load-Balancing UNIX. Proc. Europ. UNIX Systems User Group Conf. Autumn '86, Manchester, UK, Sep. 1986. Buntingford Herts, UK: EUUG, pp. 273–280

192. Barak, A., Paradise, O.G.: MOS: Scaling up UNIX. Proc. USENIX Conf. Summer '86, Atlanta, GA, Jun. 1986. Berkeley, CA: USENIX Association, pp. 414–418

193. Barak, A., Malki, D.: Distributed Light Weighted Processes in MOS. Proc. Europ. UNIX Systems User Group Conf. Autumn '88, Cascais, Portugal, Oct. 1988. Buntingford Herts, UK: EUUG, pp. 335–343

194. Drezner, Z., Barak, A.: A Probabilistic Algorithm for Scattering Information in a Multi-computer System. Computing Research Lab., Univ. of Michigan, Technical Report CRL–TR–15–84, Mar. 1984

195. Hofner, R.: The MOS Communication System. Dept. of Computer Science, Hebrew Univ. of Jerusalem, Israel, Master's thesis, Feb. 1984

196. Masel, J.: The MOS Filing System. Dept. of Computer Science, Hebrew Univ. of Jerusalem, Israel, Master's thesis, Aug. 1983

197. Shiloh, A.: Load Sharing in a Distributed Operating System. Dept. of Computer Science, Hebrew Univ. of Jerusalem, Israel, Master's thesis, Jul. 1983

Newcastle (Sect. 4.15)

198. Black, J.P., Marshall, L.F., Randell, B.: The Architecture of UNIX United. Proc. IEEE 75:5, 709–718 (May 1987)

199. Brownbridge, D.R., Marshall, L.F., Randell, B.: The Newcastle Connection or UNIXes of the World Unite! Software – Practice and Experience 12:12, 1147–1162 (Dec. 1982), also in: Shrivastava S.K. (ed.): Reliable Computer Systems. Collected Papers of the Newcastle Reliability Project, 1985. Berlin, Heidelberg, New York: Springer-Verlag, pp. 532–549

200. Linton, A., Panzieri, F.: A Communication System Supporting Large Datagrams on a Local Area Network. Computing Lab., Univ. of Newcastle upon Tyne, UK, Technical Report SRM/193, May 1984

201. Marshall, L.F., Stroud, R.J.: Remote File Systems Are Not Enough. Proc. Europ. UNIX Systems User Group Conf. Autumn '86, Manchester, UK, Sep. 1986. Buntingford Herts, UK: EUUG

202. Panzieri, F.: Software and Hardware of High Integrity Systems Projects: Final Report. Computing Lab., Univ. of Newcastle upon Tyne, UK, Technical Report ERC/9/4/2043/059 RSRE, Jun. 1982

203. Panzieri, F., Randell, B.: Interfacing UNIX to Data Communications Networks. Computing Lab., Univ. of Newcastle upon Tyne, UK, Technical Report 190, Dec. 1983

204. Panzieri, F., Shrivastava, S.K.: Rajdoot: An RPC Mechanism Supporting Orphan Detection and Killing. IEEE Transactions on Software Engineering SE–14:1, 30–37 (Jan. 1988)

205. Rushby, J., Randell, B.: A Distributed Secure System. IEEE Computer 16:7, 55–67 (Jul. 1983)

206. Shrivastava, S.K., Panzieri, F.P.: The Design of a Reliable RPC Mechanism. IEEE Transactions on Computers c–31:7, 692–697 (Jul. 1982)

207. Snow, C.R., Whitfield, H.: An Experiment with the Newcastle Connection Protocol. Software – Practice and Experience **16**:11, 1031–1043 (Nov. 1986)

PULSE (Sect. 4.16)

208. Keeffe, D., Tomlinson, G.M., Wand, I.C., Wellings, A.J.: PULSE: An Ada-Based Distributed Operating System. London: Academic Press, 1985

209. Keeffe, D., Tomlinson, G.M., Wand, I.C., Wellings, A.J.: Aspects of Portability of the UNIX Shell. Proc. IEEE Computer and Digital Techniques **132**:5, 257–264 (Sep. 1985)

210. Tomlinson, G.M., Keeffe, D., Wand, I.C., Wellings, A.J.: The PULSE Distributed File System. Software – Practice and Experience **15**:11, 1087–1101 (Nov. 1985)

211. Wellings, A.J.: Distributed Operating Systems and the Ada Programming Language. Dept. of Computer Science, Univ. of York, UK, PhD thesis, Apr. 1984

212. Wellings, A.J.: Communication between Ada Programs. Proc. Ada Applications and Environments, St. Paul, MN, Oct. 1984, pp. 143–153

213. Wellings, A.J.: The PULSE Project. Proc. ACM SIGOPS Workshop on Operating Systems in Computer Networks, Rüschlikon, Switzerland, Jan. 1985. ACM SIGOPS Operating Systems Review **19**:2, 6–40, Apr. 1985

QuickSilver (Sect. 4.17)

214. Cabrera, L.-F., Wyllie, J.: QuickSilver Distributed File Services: An Architecture for Horizontal Growth. IBM Almaden Research Center, San Jose, Technical Report RJ 5578 (56697), Jan. 1987, also in Proc. IEEE 2nd Conf. on Computer Workstations, Santa Clara, CA, 1988

215. Haskin, R., Malachi, Y., Sawdon, W., Chan, G.: Recovery Management in QuickSilver. ACM Transactions on Computer Systems **6**:1, 82–108 (Feb. 1988)

216. Theimer, M., Cabrera, L.-F., Wyllie, J.: QuickSilver Support for Access to Data in Large, Geographically Dispersed Systems. Proc. 9th IEEE Int. Conf. on Distributed Computing Systems, Newport Beach, CA, Jun. 1989. Los Alamitos, CA: IEEE Comp. Soc. Press, pp. 28–35

RHODOS (Sect. 4.18)

217. Gerrity, G.W., Goscinski, A., Indulska, J., Toomey, W., Zhu, W.: The RHODOS Distributed Operating System. Dept. of Computer Science, Univ. College, Univ. of New South Wales, Canberra, Australia, Technical Report CS 90/4, Feb. 1990

218. Gerrity, G.W., Goscinski, A., Indulska, J., Toomey, W.: Interprocess Communication in RHODOS. Dept. of Computer Science, Univ. College, Univ. of New South Wales, Canberra, Australia, Technical Report CS 90/6, Feb. 1990

219. Goscinski, A., Bearman, M.: Resource Management in Large Distributed Systems. ACM SIGOPS Operating Systems Review **24**:4, 7–25 (1990)

220. Goscinski, A., Indulska, J., Reynolds, P., Toomey, W.: The Development of the RHODOS Reliable Datagram Protocol. Dept. of Computer Science, Univ. College, Univ. of New South Wales, Canberra, Australia, Technical Report, 1991

221. Goscinski, A., Indulska, J., Reynolds, P., Toomey, W.: The Implementation of the RHODOS Reliable Datagram Protocol. Dept. of Computer Science, Univ. College, Univ. of New South Wales, Canberra, Australia, Technical Report, 1991

222. Toomey, W.: Memory Management in RHODOS. Dept. of Computer Science, Univ. College, Univ. of New South Wales, Canberra, Australia, Technical Report CS 90/19, May 1990

223. Zhu, W., Goscinski, A., Gerrity, G.W.: Process Migration in RHODOS. Dept. of Computer Science, Univ. College, Univ. of New South Wales, Canberra, Australia, Technical Report CS 90/9, Jul. 1989

224. Zhu, W., Goscinski, A.: Load Balancing in RHODOS. Dept. of Computer Science, Univ. College, Univ. of New South Wales, Canberra, Australia, Technical Report CS 90/8, Mar. 1990

Saguaro (Sect. 4.19)

225. Andrews, G.R.: Distributed Systems Research at Arizona. Proc. 2nd ACM SIGOPS Europ. Workshop on Making Distributed Systems Work, Amsterdam, the Netherlands, Sep. 1986. ACM SIGOPS Operating Systems Review **21**:1, 49–84, Jan. 1987

226. Andrews, G.R., Schlichting, R.D., Hayes, R., Purdin, T.: The Design of the Saguaro Distributed Operating System. IEEE Transactions on Software Engineering **SE–13**:1, 104–118 (Jan. 1987)

227. Hayes, R.: UTS: A Type System for Facilitating Data Communication. Dept. of Computer Science, Univ. of Arizona, AZ, PhD thesis, Aug. 1989, available as Technical Report TR 89-16

228. Purdin, T.: Enhancing File Availability in Distributed Systems (The Saguaro File System). Dept. of Computer Science, Univ. of Arizona, AZ, PhD thesis, Aug. 1987, available as Technical Report TR 87-26

229. Schlichting, R.D., Andrews, G.R., Purdin, T.: Mechanisms to Enhance File Availability in Distributed Systems. Proc. 16th Int. Symp. Fault-Tolerant Computing, Vienna, Austria, Jul. 1986, pp. 44–49

230. Schlichting, R.D., Purdin, T., Andrews, G.R.: A File Replication Facility for Berkeley UNIX. Software – Practice and Experience **17**:12, 923–940 (Dec. 1987)

231. Schlichting, R.D., Andrews, G.R., Hutchinson, N.C., Olsson, R., Peterson, L.L.: Observations on Building Distributed Languages and Systems. In: Nehmer, J. (ed.): Proc. Int. Workshop on Experiences with Distributed Systems, Kaiserslautern, Germany, 1988. Lecture Notes in Computer Science **309**. Berlin, Heidelberg, New York: Springer-Verlag, pp. 271–291

Sprite (Sect. 4.20)

232. Baker, M., Ousterhout, J.K.: Availability in the Sprite Distributed File System. Proc. 4th ACM SIGOPS Europ. Workshop on Fault Tolerance Support in Distributed Systems, Bologna, Italy, Sep. 1990. ACM SIGOPS Operating Systems Review **25**:2

233. Douglis, F., Ousterhout, J.K.: Process Migration in the Sprite Operating System. Proc. 7th IEEE Int. Conf. on Distributed Computing Systems, Berlin, Germany, Sep. 1987. Los Alamitos, CA: IEEE Comp. Soc. Press, pp. 18–25

234. Douglis, F., Ousterhout, J.K.: Process Migration in Sprite: A Status Report. IEEE Computer Soc. Techn. Committee on Operating Systems Newsletter **3**:1, 8–10 (1989)

235. Douglis, F.: Experience with Process Migration in Sprite. Proc. 1st USENIX Workshop on Experiences with Distributed and Multiprocessor Systems, Ft. Lauderdale, FL, Oct. 1989. Berkeley, CA: USENIX Association, pp. 59–72

236. Douglis, F., Ousterhout, J.K.: Log-Structured File Systems. Proc. IEEE COMPCON Conf. Spring '89, San Francisco, CA, Feb. 1989. Los Alamitos, CA: IEEE Comp. Soc. Press, pp. 124–129

237. Douglis, F.: Transparent Process Migration in the Sprite Operating System. Computer Science Division, Univ. of California, Berkeley, CA, PhD thesis, Sep. 1990, available as Technical Report UCB/CSD 90/598

238. Hartman, J.H., Ousterhout, J.K.: Performance Measurements of a Sprite Multiprocessor Kernel. Proc. USENIX Conf. Summer '90, Anaheim, CA, Jun. 1990. Berkeley, CA: USENIX Association, pp. 279–287

239. Nelson, M.N., Ousterhout, J.K.: Copy-on-Write for Sprite. Proc. USENIX Conf. Summer '88, San Francisco, CA, Jun. 1988. Berkeley, CA: USENIX Association, pp. 187–201

240. Nelson, M.N., Welch, B.B., Ousterhout, J.K.: Caching in the Sprite Network File System. ACM Transactions on Computer Systems 6:1, 134–154 (Feb. 1988)

241. Nelson, M.N.: Virtual Memory for the Sprite Operating System. Computer Science Division, Univ. of California, Berkeley, CA, Technical Report UCB/CSD 86/301, Jun. 1986

242. Nelson, M.N.: Physical Memory Management in a Network Operating System. Computer Science Division, Univ. of California, Berkeley, CA, PhD thesis, Nov. 1988, available as Univ. of California, Berkeley, CA, Technical Report UCB/CSD 88/471

243. Ousterhout, J.K., Douglis, F.: Beating the I/O Bottleneck: A Case for Log-Structured File Systems. ACM SIGOPS Operating Systems Review 23:1, 11–28 (Jan. 1989), also available as Univ. of California, Berkeley, CA, Technical Report UCB/CSD 88/467

244. Ousterhout, J.K., et al.: A Trace-Driven Analysis of the Unix 4.2 BSD File System. Proc. 10th ACM Symp. on Operating Systems Principles, Orcas Island, WA, Dec. 1985. ACM SIGOPS Operating Systems Review 19:5, pp. 15–24

245. Ousterhout, J.K., Cherenson, A.R., Douglis, F., Nelson, M.N., Welch, B.B.: The Sprite Network Operating System. IEEE Computer 21:2, 23–36 (Feb. 1988)

246. Ousterhout, J.K.: Why Aren't Operating Systems Getting Faster As Fast As Hardware. Proc. USENIX Conf. Summer '90, Anaheim, CA, Jun. 1990. Berkeley, CA: USENIX Association, pp. 247–256

247. Rosenblum, M., Ousterhout, J.K.: The LFS Storage Manager. Proc. USENIX Conf. Summer '90, Anaheim, CA, Jun. 1990. Berkeley, CA: USENIX Association, pp. 315–324

248. Welch, B.B.: The Sprite Remote Procedure Call System. Computer Science Division, Univ. of California, Berkeley, CA, Technical Report UCB/CSD 86/302, Jun. 1986

249. Welch, B.B., Ousterhout, J.K.: Prefix Tables: A Simple Mechanism for Locating Files in a Distributed Filesystem. Proc. 6th IEEE Int. Conf. on Distributed Computing Systems, Cambridge, MA, May 1986. Los Alamitos, CA: IEEE Comp. Soc. Press, pp. 184–189

250. Welch, B.B., Ousterhout, J.K.: Pseudo Devices: User-Level Extensions to the Sprite File System. Proc. USENIX Conf. Summer '88, San Francisco, CA, Jun. 1988. Berkeley, CA: USENIX Association, pp. 37–49

251. Welch, B.B.: Naming, State Management, and User-Level Extensions in the Sprite Distributed File System. Computer Science Division, Univ. of California, Berkeley, CA, PhD thesis, Feb. 1990, available as Technical Report UCB/CSD 90/567

V   (Sect. 4.21)

252. Cheriton, D.R.: Local Networking and Internetworking in the V-System. Proc. 7th ACM/IEEE Symp. on Data Communications, Oct. 1983, pp. 9–16

253. Cheriton, D.R., Zwaenepoel, W.: The Distributed V Kernel and its Performance for Diskless Workstations. Proc. 9th ACM Symp. on Operating Systems Principles, Bretton Woods, NH, Oct. 1983. ACM SIGOPS Operating Systems Review 17:5, pp. 129–140

254. Cheriton, D.R.: An Experiment using Registers for Fast Message-based Interprocess Communication. ACM SIGOPS Operating Systems Review 18:4, 12–20 (Oct. 1984)

255. Cheriton, D.R.: The V Kernel: A Software Base for Distributed Systems. IEEE Software 1:2, 19–42 (Apr. 1984)

256. Cheriton, D.R.: Uniform Access to Distributed Name Interpretation in the V-System. Proc. 4th IEEE Int. Conf. on Distributed Computing Systems, San Francisco, CA, May 1984. Los Alamitos, CA: IEEE Comp. Soc. Press, pp. 290–297

257. Cheriton, D.R., Zwaenepoel, W.: One-to-many Interprocess Communication in the V-System. Proc. ACM SIGCOMM Symp. in Communications Architectures and Protocols, Jun. 1984. ACM SIGCOMM Computer Communication Review, 14

258. Cheriton, D.R.: Distributed Process Groups in the V Kernel. ACM Transactions on Computer Systems 2:2, 77–107 (May 1985)

259. Cheriton, D.R.: Request-Response and Multicast Interprocess Communication in the V Kernel. In: Müller, G., Blanc, R.P. (eds.): Proc. Int. Seminar on Networking in Open Systems, Oberlech, Austria, Aug. 1986. Lecture Notes in Computer Science 248. Berlin, Heidelberg, New York: Springer-Verlag, pp. 296–312

260. Cheriton, D.R.: Problem-Oriented Shared Memory: A Decentralized Approach to Distributed Systems Design. Proc. 6th IEEE Int. Conf. on Distributed Computing Systems, Cambridge, MA, May 1986. Los Alamitos, CA: IEEE Comp. Soc. Press, pp. 190–197

261. Cheriton, D.R.: VMTP: A Transport Protocol for the Next Generation of Communication Systems. Proc. ACM SIGCOMM Symp. in Communications Architectures and Protocols, Stowe, VT, Aug. 1986. ACM SIGCOMM Computer Communication Review 16:3

262. Cheriton, D.R.: UIO: A Uniform I/O System Interface for Distributed Systems. ACM Transactions on Computer Systems 5:1, 12–46 (Feb. 1987)

263. Cheriton, D.R., Williamson, C.: Network Measurement of the VMTP Request-Response Protocol in the V Distributed System. Proc. ACM SIGMETRICS Conf. on Measurement and Modeling of Computer Systems, Banff, Canada, May 1987. ACM SIGMETRICS Performance Evaluation Review 15:1

264. Cheriton, D.: The V Distributed System. Communications of the ACM 31:3, 314–333 (Mar. 1988)

265. Cheriton, D.R.: Unified Management of Memory and File Caching Using the V Virtual Memory System. Dept. of Computer Science, Stanford Univ., Stanford, CA, Technical Report STAN–CS–88–1192, 1988

266. Cheriton, D.R., Mann, T.P.: Decentralizing: A Global Name Service for Efficient Fault-Tolerant Access. ACM Transactions on Computer Systems 7:2, 147–183 (May 1989)

267. Lantz, K.A., Nowicki, W.I.: Structured Graphics for Distributed Systems. ACM Transactions on Computer Systems 3:1, 23–51 (Jan. 1984)

268. Lantz, K.A., Nowicki, W.I.: Virtual Terminal Services in Workstation-Based Distributed Systems. Proc. 17th IEEE Hawaii Int. Conf. on System Sciences, Hawaii, HI, Jan. 1984. Los Alamitos, CA: IEEE Comp. Soc. Press, pp. 196–205

269. Stumm, M.: Verteilte Systeme: Eine Einführung am Beispiel V. Informatik-Spektrum **10**:5, 246–261 (Oct. 1987)

270. Theimer, M.M., Lantz, K.A., Cheriton, D.R.: Preemptable Remote Execution Facility in the V-System. Proc. 10th ACM Symp. on Operating Systems Principles, Orcas Island, WA, Dec. 1985. ACM SIGOPS Operating Systems Review **19**:5

271. Zwaenepoel, W.: Implementation and Performance of Pipes in the V-System. IEEE Transactions on Computers **c–34**:12, 1174–1178 (Dec. 1985)

WANDA (Sect. 4.22)

272. McAuley, D.: Protocol Design for High Speed Networks. Computer Lab., Cambridge, UK, PhD thesis, 1989, available as Technical Report TR-186

273. Thomson, S.E.: A Storage Service for Structured Data. Computer Lab., Cambridge, UK, PhD thesis, 1991

274. Wilson, T.D.: Increasing Performance of Storage Services. Computer Lab., Cambridge, UK, PhD thesis, 1991

Wisdom (Sect. 4.23)

275. Austin, P.B., Murray, K.A., Wellings, A.J.: The Design of Scalable Parallel Operating Systems. Dept. of Computer Science, Univ. of York, UK, Technical Report, Jan. 1989

276. Austin, P.B., Murray, K.A., Wellings, A.J.: Wisdom: A Prototype Scalable Operating System. Proc. 2nd IEEE Workshop on Experimental Distributed Systems, Huntsville, AL, Oct. 1990. Los Alamitos, CA: IEEE Comp. Soc. Press, pp. 106–112

277. Austin, P.B., Murray, K.A., Wellings, A.J.: File System Caching in Large Point-to-Point Networks. Dept. of Computer Science, Univ. of York, UK, Technical Report, Sep. 1990

278. Murray, K.A., Wellings, A.J.: Issues in the Design and Implementation of a Distributed Operating System for a Network of Transputers. Microprocessing and Microprogramming, Proc. of EUROMICRO '88, pp. 169–178 (Sep. 1988)

279. Murray, K.A., Wellings, A.J.: Wisdom: A Distributed Operating System for Transputers. Computer Systems: Science and Engineering **5**:1, 13–20 (Jan. 1990)

280. Murray, K.A.: Wisdom: The Foundation of a Scalable Parallel Operating System. Dept. of Computer Science, Univ. of York, UK, Technical Report, Feb. 1990

Alpha (Sect. 5.1)

281. Jensen, E.D.: Alpha: A Non-Proprietary Operating System for Mission-Critical Real-Time Distributed Systems. Concurrent Computer Corp., Technical Report 89121, Dec. 1989

282. Jensen, E.D., Northcutt, J.D.: Alpha: A Non-Proprietary Operating System for Mission-Critical Real-Time Distributed Systems. Proc. 2nd IEEE Workshop on Experimental Distributed Systems, Huntsville, AL, Oct. 1990. Los Alamitos, CA: IEEE Comp. Soc. Press

283. Northcutt, J.D.: Mechanisms for Reliable Distributed Real-Time Operating Systems: The Alpha Kernel. New York: Academic Press, 1987

284. Reynolds, F.D., Northcutt, J.D., Jensen, E.D., Clark, R.K., Shipman, S.E., Dasarathy, B., Maynard, D.P.: Threads: A Programming Construct for Reliable Distributed Computing. Int. Journal of Mini and Microcomputer **12**:3 (1990)

Amoeba (Sect. 5.2)

285. Baalbergen, E.H.: Parallel and Distributed Compilations in Loosely-Coupled Systems. Proc. 1st Workshop on Large-Grained Parallelism, Providence, RI, Oct. 1986

286. Baalbergen, E.H.: Design and Implementation of Parallel Make. Computing Systems **1**:2, 135–158 (1988)

287. Baalbergen, E.H., Verstoep, K., Tanenbaum, A.S.: On the Design of the Amoeba Configuration Manager. Proc. 2nd ACM SIGSOFT Int. Workshop on Software Configuration Management, Nov. 1989. ACM SIGSOFT Engineering Notes, **17**

288. Bal, H.E., Renesse, R. van, Tanenbaum, A.S.: Implementing Distributed Algorithms using Remote Procedure Call. Proc. AFIPS Nat. Computer Conf., 1987, pp. 499–505

289. Bal, H.E., Tanenbaum, A.S.: Distributed Programming with Shared Data. Proc. IEEE Int. Conf. on Computer Languages, 1988. Los Alamitos, CA: IEEE Comp. Soc. Press, pp. 82–91

290. Bal, H.E., Kaashoek, M.F., Tanenbaum, A.S.: A Distributed Implementation of the Shared Data-object Model. Proc. 1st USENIX Workshop on Experiences with Distributed and Multiprocessor Systems, Ft. Lauderdale, FL, Oct. 1989. Berkeley, CA: USENIX Association, pp. 1–19

291. Bal, H.E., Kaashoek, M.F., Tanenbaum, A.S.: Experience with Distributed Programming in Orca. Proc. IEEE Int. Conf. on Computer Languages, 1990. Los Alamitos, CA: IEEE Comp. Soc. Press

292. Kaashoek, M.F., Tanenbaum, A.S., Flynn Hummel, S., Bal, H.E.: An Efficient Reliable Broadcast Protocol. ACM SIGOPS Operating Systems Review **23**:4, 5–19 (Oct. 1989)

293. Kaashoek, M.F., Tanenbaum, A.S.: Group Communication in the Amoeba Distributed Operating System. Proc. 11th IEEE Int. Conf. on Distributed Computing Systems, Arlington, TX, May 1991. Los Alamitos, CA: IEEE Comp. Soc. Press

294. Mullender, S.J., Rossum, G. van, Tanenbaum, A.S., Renesse, R. van, Staveren, H. van: Amoeba: A Distributed Operating System for the 1990s. IEEE Computer **23**:5, 44–53 (May 1990)

295. Renesse, R. van, Tanenbaum, A.S., Wilschut, A: The Design of a High-Performance File Server. Proc. 9th IEEE Int. Conf. on Distributed Computing Systems, Newport Beach, CA, Jun. 1989. Los Alamitos, CA: IEEE Comp. Soc. Press, pp. 22–27

296. Renesse, R. van, Staveren, H. van, Tanenbaum, A.S.: Performance of the Amoeba Distributed Operating System. Software – Practice and Experience **19**:3, 223–234 (Mar. 1989)

297. Renesse, R. van, Tanenbaum, A.S., Staveren, H., Hall, J.: Connecting RPC-Based Distributed Systems using Wide-Area Networks. Proc. 7th IEEE Int. Conf. on Distributed Computing Systems, Berlin, Germany, Sep. 1987. Los Alamitos, CA: IEEE Comp. Soc. Press, pp. 28–34

298. Rossum, G. van: AIL A Class-Oriented Stub Generator for Amoeba. Proc. 1st USENIX Workshop on Experiences with Distributed and Multiprocessor Systems, Ft. Lauderdale, FL, Oct. 1989. Berkeley, CA: USENIX Association

299. Tanenbaum, A.S., Mullender, S.J., Renesse, R.van: Using Sparse Capabilities in a Distributed Operating Systems. Proc. 6th IEEE Int. Conf. on Distributed Computing Systems, Cambridge, MA, May 1986. Los Alamitos, CA: IEEE Comp. Soc. Press, pp. 558–563

300. Tanenbaum, A.S., Renesse, R. van, Staveren, H. van, Sharp, G.J., Mullender, S.J., Jansen, J., Rossum, G. van: Experiences with the Amoeba Distributed Operating System. Communications of the ACM **33**:12, 46–63 (Dec. 1990)

Argus (Sect. 5.3)

301. Greif, I., Seliger, R., Weihl, W.: A Case Study of CES: A Distributed Collaborative Editing System Implementation in Argus. Programming Methodology Group Memo 55 (Apr. 1987), MIT Lab. for Computer Science, Cambridge, MA

302. Ladin, R., Liskov, B., Shrira, L.: A Technique for Constructing Highly Available Services. Algorithmica **3**, 393–420 (1988)

303. Liskov, B., Snyder, A., Atkinson, R.R., Schaffert, J.C.: Abstraction mechanisms in CLU. Communications of the ACM **20**:8, 564–576 (Aug. 1977)

304. Liskov, B.: The Argus Language and System. In: Paul, M., Siegert, H.-J. (eds.): Proc. Advanced Course on Distributed Systems – Methods and Tools for Specification, Munich, Germany, Apr. 1985. Lecture Notes in Computer Science **190**. Berlin, Heidelberg, New York: Springer-Verlag, pp. 343–430

305. Liskov, B., Scheifler, R.W.: Guardians and Actions: Linguistic Support for Robust, Distributed Programs. ACM Transactions on Programming Languages and Systems **5**:3, 381–404 (Jul. 1983)

306. Liskov, B.: Overview of the Argus language and system. Programming Methodology Group Memo 40 (Feb. 1984), MIT Lab. for Computer Science, Cambridge, MA

307. Liskov, B., Ladin, R.: Highly-Available Distributed Services and Fault-Tolerant Distributed Garbage Collection. Proc. 5th ACM Symp. on Principles of Distributed Computing, Vancouver, Canada, Aug. 1986, pp. 29–39

308. Liskov, B., Curtis, D., Johnson, P., Scheifler, R.: Implementation of Argus. Proc. 11th ACM Symp. on Operating Systems Principles, Austin, TX, Nov. 1987. ACM SIGOPS Operating Systems Review **21**:5, pp. 111–122

309. Liskov, B., Scheifler, R., Walker, E., Weihl, W.: Orphan Detection. Programming Methodology Group Memo 53 (Jul. 1987), MIT Lab. for Computer Science, Cambridge, MA

310. Liskov, B., et al.: Argus Reference Manual. MIT Lab. for Computer Science, Cambridge, MA, Technical Report MIT/LCS/TR–400, 1987

311. Liskov, B.: Distributed Programming in Argus. Communications of the ACM **31**:3, 300–312 (Mar. 1988)

312. Walker, E.W.: Orphan Detection in the Argus System. MIT Lab. for Computer Science, Cambridge, MA, Technical Report MIT/LCS/TR 326, Jun. 1984

313. Weihl, W., Liskov, B.: Implementation of Resilient, Atomic Data Types. ACM Transactions on Programming Languages and Systems **7**:2, 244–269 (Apr. 1985)

BirliX (Sect. 5.4)

314. Härtig, H., Kühnhauser, W., Lux, W., Streich, H., Goos, G.: Structure of the BirliX Operating System. Proc. Europ. UNIX Systems User Group Conf. Autumn '86, Manchester, UK, Sep. 1986. Buntingford Herts, UK: EUUG, pp. 433–449

315. Härtig, H., Kühnhauser, W., Lux, W., Streich, H.: Distribution in the BirliX Operating System. Proc. 9th NTG/GI Conf. Architektur und Betrieb von Rechensystemen, Stuttgart, Germany, Mar. 1986. NTG-Fachberichte **92**. Berlin, Offenbach: VDE-Verlag, pp. 345–357

316. Härtig, H., Kühnhauser, W., Lux, W., Streich, H., Goos, G.: Distribution and Recovery in the BirliX Operating System. In: Gerner, N., Spaniol, O. (eds.): Proc. 5th GI/NTG Conf. – Communication in Distributed Systems, Aachen, Germany, Feb. 1987. Informatik-Fachberichte **130**. Berlin, Heidelberg, New York: Springer-Verlag, pp. 190–201

317. Härtig, H., Kühnhauser, W., Lux, W., Reck, W., Streich, H.: The BirliX Operating System. GMD, St. Augustin, Germany, Technical Report, Jun. 1988

318. Härtig, H., Kühnhauser, W., Kowalski, O., Lux, W., Reck, W., Streich, H., Goos, G.: The Architecture of the BirliX Operating System. In: Müller-Stoy, P. (ed.): Proc. 11th ITG/GI Conf. Architektur von Rechensystemen, Munich, Germany, Mar. 1990. Berlin, Offenbach: VDE-Verlag, pp. 339–354

319. Kowalski, O.C., Härtig, H.: Protection in the BirliX Operating System. Proc. 10th IEEE Int. Conf. on Distributed Computing Systems, Paris, France, May 1990. Los Alamitos, CA: IEEE Comp. Soc. Press, pp. 160–166

CHORUS (Sect. 5.5)

320. Abrossimov, V., Rozier, M., Shapiro, M.: Generic Virtual Memory Management for Operating System Kernels. Proc. 12th ACM Symp. on Operating Systems Principles, The Wigwam Litchfield Park, AZ, Dec. 1989. ACM SIGOPS Operating Systems Review **23**:5, pp. 123–136

321. Abrossimov, V., Rozier, M., Gien, M.: Virtual Memory Management in Chorus. In: Schröder-Preikschat, W., Zimmer, W. (eds.): Proc. Europ. Workshop on Progress in Distributed Operating Systems and Distributed Systems Management, Berlin, Germany, Apr. 1989. Lecture Notes in Computer Science **433**. Berlin, Heidelberg, New York: Springer-Verlag

322. Armand, F., Gien, M., Guillemont, M., Leonard, P.: Towards a Distributed UNIX System – The CHORUS Approach. Proc. Europ. UNIX Systems User Group Conf. Autumn '86, Manchester, UK, Sep. 1986. Buntingford Herts, UK: EUUG, pp. 413–431

323. Armand, F., Gien, M., Hermann, F., Rozier, M.: Revolution 89 or "Distributed UNIX Brings it Back to its Original Virtues". Proc. 1st USENIX Workshop on Experiences with Distributed and Multiprocessor Systems, Ft. Lauderdale, FL, Oct. 1989. Berkeley, CA: USENIX Association, pp. 153–174

324. Armand, F., Hermann, F., Lipkis, J., Rozier, M.: Multi-Threaded Process in Chorus/MIX. Proc. Europ. UNIX Systems User Group Conf. Spring '90, Munich, Germany, Apr. 1990. Buntingford Herts, UK: EUUG, pp. 1–13

325. Banino, J.S., Caristan, A., Guillemont, M., Morisset, G., Zimmermann, H.: CHORUS: An Architecture for Distributed Systems. INRIA, France, Technical Report 42, Nov. 1980

326. Banino, J.S., Fabre, J.C.: Distributed Coupled Actors: A CHORUS Proposal for Reliability. Proc. 3rd IEEE Int. Conf. on Distributed Computing Systems, Ft. Lauderdale, FL, Oct. 1982. Los Alamitos, CA: IEEE Comp. Soc. Press, pp. 128–134

327. Banino, J.S.: Parallelism and Fault-Tolerance in the CHORUS Distributed System. Proc. Int. Workshop on Modelling and Performance Evaluation of Parallel Systems, Grenoble, France, Dec. 1984

328. Banino, J.S., Fabre, J.C., Guillemont, M., Morisset, G., Rozier, M.: Some Fault Tolerant Aspects of the CHORUS Distributed System. Proc. 5th IEEE Int. Conf. on Distributed Computing Systems, Denver, CO, May 1985. Los Alamitos, CA: IEEE Comp. Soc. Press, pp. 430–437

329. Banino, J.S., Morisset, G., Rozier, M.: Controlling Distributed Processing with CHORUS Activity Messages. Proc. 18th IEEE Hawaii Int. Conf. on System Sciences, Hawaii, HI, Jan. 1985. Los Alamitos, CA: IEEE Comp. Soc. Press

330. Guillemont, M.: The CHORUS Distributed Computing System: Design and Implementation. In: Ravasio, P.C., Hopkins, G., Naffah, N. (eds.): Proc. Int. Symp. on Local Computer Networks, Florence, Italy, Apr. 1982. Amsterdam, New York: North-Holland, pp. 207–223

331. Guillemont, M.: Chorus: A Support for Distributed and Reconfigurable ADA Software. Proc. ESA Workshop on Communication Networks and Distributed Operating Systems within the Space Environment, Oct. 1989

332. Herrmann, F., Armand, F., Rosier, M., Gien, M., Abrossimov, V., Boule, I., Guillemont, M., Léonard, P., Langlois, S., Neuhauser, W.: CHORUS, a new Technology for Building UNIX-Systems. Proc. Europ. UNIX Systems User Group Conf. Autumn '88, Cascais, Portugal, Oct. 1988. Buntingford Herts, UK: EUUG, pp. 1–18

333. Herrmann, F., Philippe, L.: Multicomputers UNIX based on CHORUS. In: Bode, A. (ed.): Proc. 2nd Europ. Distributed Memory Computing Conf., Munich, Germany, Apr. 1991. Lecture Notes in Computer Science 487. Berlin, Heidelberg, New York: Springer-Verlag, pp. 440–449

334. Rozier, M., Legatheaux, J.M.: The CHORUS Distributed Operating System: Some Design Issues. In: Paker, Y., Banatre, J.-P., Bozyigit, M. (eds.): Distributed Operating Systems: Theory and Practice, NATO ASI series Vol. F28, 1987. Berlin, Heidelberg, New York: Springer-Verlag, pp. 261–289

335. Rozier, M., et al.: The CHORUS Distributed Operating System. Computing Systems 1:4, 305–370 (1988)

336. Zimmermann, H., Banino, J.S., Caristan, A., Guillemont, M., Morisset, G.: Basic Concepts for the support of Distributed Systems: The CHORUS approach. Proc. 2nd IEEE Int. Conf. on Distributed Computing Systems, 1981. Los Alamitos, CA: IEEE Comp. Soc. Press, pp. 60–66

337. Zimmermann, H., Guillemont, M., Morisset, G., Banino, J.S.: CHORUS: A Communication and Processing Architecture for Distributed Systems. INRIA Research, Technical Report 328, Sep. 1984

Clouds (Sect. 5.6)

338. Ahamad, M., Dasgupta, P.: Parallel Execution Threads: An Approach to Atomic Actions. School of Information and Computer Science, Georgia Inst. of Technology, Atlanta, GA, Technical Report GIT-ICS-87/16, 1987

339. Ahamad, M., Ammar, M.H., Bernabéau, J., Khalidi, M.Y.A.: A Multicast Scheme for Locating Objects in a Distributed Operating System. School of Information and Computer Science, Georgia Inst. of Technology, Atlanta, GA, Technical Report GIT-ICS-87/01, 1987

340. Allchin, J.E.: An Architecture for Reliable Decentralized Systems. School of Information and Computer Science, Georgia Inst. of Technology, Atlanta, GA, PhD thesis, Technical Report GIT-ICS-83/23, 1983

341. Bernabéau-Aubán, J.M., Hutto, P.W., Khalidi, M.Y.A.: The Architecture of the Ra Kernel. School of Information and Computer Science, Georgia Inst. of Technology, Atlanta, GA, Technical Report GIT-ICS-87/35, 1987

342. Bernabéau-Aubán, J.M., Hutto, P.W., Yousef, M., Khalidi, M.Y.A., Ahamad, M., Appelbe, W.F., Dasgupta, P., LeBlanc, R.J., Ramachandran, U.: Clouds – A Distributed, Object-Based Operating System, Architecture and Kernel Implementation. Proc. Europ. UNIX Systems User Group Conf. Autumn '88, Cascais, Portugal, Oct. 1988. Buntingford Herts, UK: EUUG, pp. 25–37

343. Chen, R.C., Dasgupta, P.: Consistency-Preserving Threads: Yet Another Approach to Atomic Programming. School of Information and Computer Science, Georgia Inst. of Technology, Atlanta, GA, Technical Report GIT-ICS-87/43, 1987

344. Chen, R.C., Dasgupta, P.: Implementing Consistency Control Mechanisms in the Clouds Distributed Operating System. Proc. 11th IEEE Int. Conf. on Distributed Computing Systems, Arlington, TX, May 1991. Los Alamitos, CA: IEEE Comp. Soc. Press

345. Dasgupta, P., LeBlanc, R.J., Spafford, E.: The Clouds Project: Design and Implementation of a Fault-Tolerant Distributed Operating System. School of Information and Computer Science, Georgia Inst. of Technology, Atlanta, GA, Technical Report GIT-ICS-85/29, 1985

346. Dasgupta, P., LeBlanc, R.J., Appelbe, W.F.: The Clouds Distributed Operating System. Proc. 8th IEEE Int. Conf. on Distributed Computing Systems, San Jose, CA, Jun. 1988. Los Alamitos, CA: IEEE Comp. Soc. Press, pp. 1–9

347. Dasgupta, P., Chen, R.C., Menon, S., Pearson, M.P., Ananthanarayan, R., Ramachandran, U., Ahamad, M., LeBlanc, R.J., Appelbe, W.F., Bernabéau, J.M., Hutto, P.W., Khalidi, M.Y.A., Wilkenloh, C.: The Design and Implementation of the Clouds Distributed Operating System. Computing Systems 3:1, 11–46 (Winter 1990)

348. Spafford, E.H.: Kernel Structures for a Distributed Operating System. School of Information and Computer Science, Georgia Inst. of Technology, Atlanta, GA, PhD thesis, Technical Report GIT-ICS-86/16, 1986

349. Spafford, E.H.: Object Operation Invocation in Clouds. School of Information and Computer Science, Georgia Inst. of Technology, Atlanta, GA, Technical Report GIT-ICS-87/14, 1987

350. Pitts, D.V., Dasgupta, P.: Object Memory and Storage Management in the Clouds Kernel. Proc. 8th IEEE Int. Conf. on Distributed Computing Systems, San Jose, CA, Jun. 1988. Los Alamitos, CA: IEEE Comp. Soc. Press, pp. 10–17

351. Wilkes, C.T.: Preliminary Aeolus Reference Manual. School of Information and Computer Science, Georgia Inst. of Technology, Atlanta, GA, Technical Report GIT-ICS-85/07, 1985

352. Wilkes, C.T., LeBlanc, R.J.: Rationale for the Design of Aeolus: A Systems Programming Language for an Action/Object System. School of Information and Computer Science,

Georgia Inst. of Technology, Atlanta, GA, Technical Report GIT-ICS-86/12, 1986

353. Wilkes, C.T., LeBlanc, R.J.: Programming Language Features for Resilience and Availability. Proc. 2nd Workshop on Large-Grained Parallelism, Hidden Valley, PA, Oct. 1987, pp. 90–92

354. Wilkes, C.T.: Programming Methodologies for Resilience and Availability. School of Information and Computer Science, Georgia Inst. of Technology, Atlanta, GA, PhD thesis, Technical Report GIT-ICS-87/32, 1987

Cronus   (Sect. 5.7)

355. Dean, M.A., Sands, R.E., Schantz, R.E.: Canonical Data Representation in the Cronus Distributed Operating System. Proc. IEEE INFOCOM, San Francisco, CA, Apr. 1987. Los Alamitos, CA: IEEE Comp. Soc. Press, pp. 814–819

356. Gurwitz, R.F., Dean, M.A., Schantz, R.E.: Programming Support in the Cronus Distributed Operating System. Proc. 6th IEEE Int. Conf. on Distributed Computing Systems, Cambridge, MA, May 1986. Los Alamitos, CA: IEEE Comp. Soc. Press, pp. 486–493

357. Schantz, R.E., Thomas, R.H., Bono, G.: The Architecture of the Cronus Distributed Operating System. Proc. 6th IEEE Int. Conf. on Distributed Computing Systems, Cambridge, MA, May 1986. Los Alamitos, CA: IEEE Comp. Soc. Press, pp. 250–259

Cosmos   (Sect. 5.8)

358. Blair, G.S., Nicol, J.R., Yip, C.K.: A Functional Model of Distributed Computing. Dept. of Computer Science, Univ. of Lancaster, Lancaster, UK, Technical Report CS-DC-1-86, 1986

359. Blair, G.S., Mariani, J.A., Nicol, J.R.: COSMOS – A Nucleus for a Program Support Environment. Dept. of Computer Science, Univ. of Lancaster, Lancaster, UK, Technical Report CS-SE-1-86, 1986

360. Blair, G.S., Nicol, J.R., Walpole, J.: An Overview of the COSMOS Distributed Programming Environment Project. Dept. of Computer Science, Univ. of Lancaster, Lancaster, UK, Technical Report CS-DC-4-87, 1987

361. Blair, G.S., Malik, J., Nicol, J.R., Walpole, J.: Design Issues in the COSMOS Distributed Operating System. Proc. 3rd ACM SIGOPS Europ. Workshop, Cambridge, UK, Sep. 1988

362. Blair, G.S., Malik, J., Nicol, J.R., Walpole, J.: A Synthesis of Object Oriented and Functional Ideas in the Design of a Distributed Software Engineering Environment. Software Engineering Journal 5:3, 193–204 (May 1990)

363. Nicol, J.R., Blair, G.S., Walpole, J.: Operating System Design: Towards a Holistic Approach. ACM SIGOPS Operating Systems Review 21:1, 11–19 (Jan. 1987)

364. Nicol, J.R., Blair, G.S., Shepherd, W.D., Walpole, J.: An Approach to Multiple Copy Update Based on Immutability. Distributed Processing, pp. 537–550 (1988)

365. Nicol, J.R., Blair, G.S., Walpole, J.: A Model to Support Consistency and Availability in Distributed Systems Architectures. Proc. IEEE Workshop on Future Trends of Distributed Computing Systems in the '90s, Hong Kong, Sep. 1988. Los Alamitos, CA: IEEE Comp. Soc. Press, pp. 418–425

366. Nicol, J.R., Blair, G.S., Walpole, J., Malik, J.: COSMOS: an Architecture for a Distributed Programming Environment. Computer Communications, Butterworth Scientific 12:3, 147–157 (Jun. 1989)

367. Walpole, J., Blair, G.S., Malik, J., Nicol, J.R.: A Unifying Model for Consistent Distributed Software Development Environments. Proc. ACM SIGSOFT/SIGPLAN Software Engineering Symp. on Practical Software Development Environments, Boston, MA, Nov. 1988, pp. 183–190

368. Walpole, J., Barber, A., Blair, G.S., Nicol, J.R.: Software Development Environment Transactions: Their Implementation and Use in Cosmos. Proc. 23rd IEEE Hawaii Int. Conf. on System Sciences, Hawaii, HI, Jan. 1990. Los Alamitos, CA: IEEE Comp. Soc. Press, pp. 493–502

Eden   (Sect. 5.9)

369. Almes, G.T., Black, A.P., Bunje, C., Wiebe, D.: Edmas: A Locally Distributed Mail System. Proc. 7th IEEE Int. Conf. on Software Engineering, Orlando, FL, Mar. 1984. Los Alamitos, CA: IEEE Comp. Soc. Press, pp. 56–66

370. Almes, G.T., Black, A.P., Lazowska, E.D., Noe, J.D.: The Eden System: A Technical Review. IEEE Transactions on Software Engineering SE–11:1, 43–58 (Jan. 1985)

371. Banawan, S.A.: An Evaluation of Load Sharing in Locally Distributed Systems. Dept. of Computer Science, Univ. of Washington, Seattle, WA, Technical Report 87-08-02, Aug. 1987

372. Black, A.P.: An Asymetric Stream Communication System. Proc. 9th ACM Symp. on Operating Systems Principles, Bretton Woods, NH, Oct. 1983. ACM SIGOPS Operating Systems Review 17:5, pp. 4–10

373. Black, A.P., Hutchinson, N.C., McCord, B.C., Raj, R.K.: EPL Programmers Guide, Version 3.0. Dept. of Computer Science, Univ. of Washington, Seattle, WA, Technical Report, Jun. 1984

374. Black, A.P.: Supporting Distributed Applications: Experience with Eden. ACM SIGOPS Operating Systems Review 19:5, 181–193 (Dec. 1985)

375. Jessop, W.H., Jacobson, D.M., Noe, J.D., Baer, J.-L., Pu, C.: The Eden Transaction Based File System. Proc. 2nd IEEE Symp. on Reliability in Distributed Software and Database Systems, Jul. 1982. Los Alamitos, CA: IEEE Comp. Soc. Press, pp. 163–169

376. Lazowska, E.D., Levy, H.M., Almes, G.T., Fischer, M.J., Fowler, R.J., Festal, S.C.: The Architecture of the Eden System. Proc. 8th ACM Symp. on Operating Systems Principles, Asilomar, CA, Dec. 1981. ACM SIGOPS Operating Systems Review 15:5, pp. 148–159

377. Noe, J.D., Proudfoot, A., Pu, C.: Replication in Distributed Systems: The Eden Experience. Proc. 1st ACM/IEEE AFIPS Fall Joint Computer Conf., Dallas, TX, Nov. 1986, pp. 1197–1209

378. Noe, J.D., Wagner, D.: Measured Performance of Time Interval Concurrency Control Techniques. Proc. 13th Int. Conf. on Very Large Data Bases, Brighton, UK, Sep. 1987. Los Altos, CA: Morgan Kaufmann Publ. Inc., pp. 359–368

379. Pu, C., Noe, J.D., Proudfoot, A.: Regeneration of Replicated Objects: A Technique and its Eden Implementation. Proc. 2nd IEEE Int. Conf. on Data Engineering, Feb. 1986. Los Alamitos, CA: IEEE Comp. Soc. Press, pp. 175–187

380. Pu, C.: Replication and Nested Transactions in the Eden Distributed System. Dept. of Computer Science, Univ. of Washington, Seattle, WA, PhD thesis, 1986

381. Pu, C., Noe, J.D.: Design and Implementation of Nested Transactions in Eden. Proc. 6th IEEE Symp. on Reliability in Distributed Software and Database Systems, Williamsburg, VA, Mar. 1987. Los Alamitos, CA: IEEE Comp. Soc. Press, pp. 126–136

382. Pu, C., Noe, J.D., Proudfoot, A.: Regeneration of Replicated Objects: A Technique and Its Eden Implementation. IEEE Transactions on Software Engineering SE–14:7, 936–945 (Jul. 1988)

Gothic (Sect. 5.10)

383. Banâtre, J.P., Banâtre, M., Ployette, F.: The Concept of Multifonction: A General Structuring Tool for Distributed Operating System. Proc. 6th ACM Symp. on Principles of Distributed Computing, Cambridge, MA, May 1986, pp. 478–485

384. Banâtre, J.P., Banâtre, M., Muller, G.: Ensuring data security and integrity with a fast stable storage. Proc. 4th IEEE Int. Conf. on Data Engineering, Los Angeles, CA, Feb. 1988. Los Alamitos, CA: IEEE Comp. Soc. Press, pp. 285–293

385. Banâtre, J.P., Banâtre, M., Morin, C.: Implementing Atomic Rendezvous within a Transactional Framework. Proc. 8th IEEE Symp. on Reliability in Distributed Software and Database Systems, Seattle, WA, Oct. 1989. Los Alamitos, CA: IEEE Comp. Soc. Press, pp. 119–128

386. Banâtre, J.P., Banâtre, M.: Systèmes distribués: concepts et expérience du système Gothic. Paris: InterEditions, 1990

387. Issarny, V.: Le traitement d'exceptions-aspects théoriques et pratiques. Technical Report 1118, INRIA, France, Nov. 1989

388. Issarny, V.: An Exception Handling mechanism for Communicating Sequential Processes. Proc. of Computer Systems and Software Engineering, Tel Aviv, Israel, May 1990

389. Michel, B.: Conception et réalisation de la mémoire virtuelle de GOTHIC. INRIA, France, PhD thesis, Sep. 1989

390. Morin, C.: Building a Reliable Communication System Using High Speed Stable Storage. Proc. 5th Int. Symp. on Computer and Information Sciences, Cappadocia, Turquey, Nov. 1990

391. Rochat, B.: Implementation of a Shared Memory System on a Loosely Coupled Multiprocessor. Proc. 5th Distributed Memory Computing Conference, Charleston, AL, Apr. 1990. Los Alamitos, CA: IEEE Comp. Soc. Press

Guide (Sect. 5.11)

392. Balter, R., Decouchant, D., Duda, A., Freyssinet, A., Krakowiak, S., Meysembourg, M., Riveill, M., Roisin, C., Pina, X. Roussetde, Scioville, R., Vandôme, G.: Experience in Object-Based Computation in the Guide Operating System. Proc. 2nd IEEE Workshop on Workstation Operating Systems, Pacific Grove, CA, Sep. 1989. Los Alamitos, CA: IEEE Comp. Soc. Press

393. Decouchant, D., Krakowiak, S., Meysembourg, M., Riveill, M., Pina, X. Roussetde: A Synchronization Mechanism for Typed Objects in a Distributed System. Proc. ACM SIGPLAN Workshop on Object-based Concurrent Programming, Sep. 1988. ACM SIGPLAN Notices, 23, 1989

394. Decouchant, D., Duda, A., Freyssinet, A., Paire, E., Riveill, M., Pina, X. Roussetde, Vandôme, G.: Guide: An Implementation of the Comandos Object-Oriented Distributed System Architecture on UNIX. Proc. Europ. UNIX Systems User Group Conf. Autumn '88, Cascais, Portugal, Oct. 1988. Buntingford Herts, UK: EUUG

395. Decouchant, D., Finn, E., Horn, C., Krakowiak, S., Riveill, M.: Experience with Implementing and Using an Object-Oriented Distributed System. Proc. 1st USENIX Workshop on Experiences with Distributed and Multiprocessor Systems, Ft. Lauderdale, FL, Oct. 1989. Berkeley, CA: USENIX Association

396. Decouchant, D., Krakowiak, S., Riveill, M.: Management of Concurrent Distributed Computation in an Object-Oriented System. Proc. 3rd Workshop on Large-Grained Parallelism, Oct. 1989

397. Decouchant, D., Paire, E., Riveill, M.: Efficient Implementation of Low-Level Synchronization Primitives in the UNIX-Based Guide Kernel. Proc. Europ. UNIX Systems User Group Conf. Autumn '89, Vienna, Austria, Sep. 1989. Buntingford Herts, UK: EUUG

398. Decouchant, D., Duda, A.: Remote Execution and Communication in Guide, an Object-Oriented Distributed System. Proc. 2nd IEEE Workshop on Experimental Distributed Systems, Huntsville, AL, Oct. 1990. Los Alamitos, CA: IEEE Comp. Soc. Press

399. Krakowiak, S., Meysembourg, M., Nguyen, H., Riveill, M., Roisin, C., Pina, X. Roussetde: Design and Implementation of an Object-Oriented, Strongly Typed Language for Distributed Applications. Journal of Object-Oriented Programming 3:3 (Sep. 1990)

400. Normand, V.: A Practical Framework for Interactive Applications in Guide, an Object-Oriented Distributed System. Proc. TOOLS '90 Conf., Jun. 1990

Gutenberg (Sect. 5.12)

401. Chrysanthis, P.K., Ramamritham, K., Stemple, D.W., Vinter, S.T.: The Gutenberg Operating System Kernel. Proc. 1st ACM/IEEE Fall Joint Computer Conf., Dallas, TX, Nov. 1986, pp. 1159–1168

402. Chrysanthis, P.K., Kodavalla, H., Ramamritham, K., Stemple, D.W.: The Gutenberg Prototype Operating System (Version 2.0). Dept. of Computer Science, Univ. of Massachusetts, Amherst, MA, Technical Report COINS TR-88-25, Mar. 1988

403. Chrysanthis, P.K., Ramamritham, K., Stemple, D.W.: The Gutenberg Experiment: Testing Design Principles as Hypotheses. Dept. of Computer Science, Univ. of Massachusetts, Amherst, MA, Technical Report COINS TR-89-36, Sep. 1989

404. Chrysanthis, P.K., Stemple, D.W., Ramamritham, K.: A Logically Distributed Approach for Structuring Office Systems. Proc. Conf. of Office Information Systems, Cambridge, MA, Apr. 1990, pp. 12–21

405. Ramamritham, K., Stemple, D.W., Vinter, S.T.: Primitives for Accessing Protected Objects. Proc. 3rd IEEE Symp. on Reliability in Distributed Software and Database Systems, Clearwater Beach, FL, Oct. 1983. Los Alamitos, CA: IEEE Comp. Soc. Press, pp. 114–121

406. Ramamritham, K., Stemple, D.W., Vinter, S.T.: Decentralized Access Control in a Distributed System. Proc. 5th IEEE Int. Conf. on Distributed Computing Systems, Denver, CO, May 1985. Los Alamitos, CA: IEEE Comp. Soc. Press, pp. 524–532

407. Ramamritham, K., Briggs, D., Stemple, D.W., Vinter, S.T.: Privilege Transfer and Revocation in a Port-Based System. IEEE Transactions on Software Engineering SE–12:5, 635–648 (May 1986)

408. Stemple, D.W., Ramamritham, K., Vinter, S.T., Sheard, T.: Operating System Support for Abstract Database Types. Proc. 2nd Int. Conf. on Databases, Sep. 1983, pp. 179–195

409. Stemple, D.W., Vinter, S.T., Ramamritham, K.: Functional Addressing in Gutenberg: Interprocess Communication Without Process Identifiers. IEEE Transactions on Software Engineering SE–12:11, 1056–1066 (Dec. 1986)

410. Vinter, S.T., Ramamritham, K., Stemple, D.W.: Protecting Objects through the Use of Ports. Proc. 2nd Phoenix Conf. on Computers and Communication, Mar. 1983. Los Alamitos, CA: IEEE Comp. Soc. Press, pp. 399–404

411. Vinter, S.T.: A Protection Oriented Distributed Kernel. Dept. of Computer Science, Univ. of Massachusetts, MA, PhD thesis, Aug. 1985

412. Vinter, S.T., Ramamritham, K., Stemple, D.W.: Recoverable Communicating Actions. Proc. 6th IEEE Int. Conf. on Distributed Computing Systems, Cambridge, MA, May 1986. Los Alamitos, CA: IEEE Comp. Soc. Press, pp. 242–249

MARUTI (Sect. 5.13)

413. Agrawala, A., Levi, S.: Objects Architecture for Real-Time Operating Systems. Proc. IEEE Workshop on Real-Time Operating Systems, Cambridge, MA, Jul. 1987. Los Alamitos, CA: IEEE Comp. Soc. Press, pp. 142–148

414. Agrawala, A., Tripathi, S., Gudmundsson, O., Mosse, D., Ko, K.: MARUTI: A Platform for Hard Real-Time Applications. Proc. Workshop on Operating Systems for Mission Critical Computing, Sep. 1989

415. Agrawala, A., Tripathi, S., Carson, S., Levi, S.: The MARUTI Hard Real-Time Operating System. ACM SIGOPS Operating Systems Review 23:3 (Jul. 1989)

416. Agrawala, A., Levi, S.: Real Time System Design. New York: McGraw-Hill, 1990

417. Gudmundsson, O., Mosse, D., Agrawala, A.K., Tripathi, S.K.: MARUTI: A Hard Real-Time Operating System. Proc. 2nd IEEE Workshop on Experimental Distributed Systems, Huntsville, AL, Oct. 1990. Los Alamitos, CA: IEEE Comp. Soc. Press

418. Levi, S., Mosse, D., Agrawala, A.: Allocation of Real-Time Computations under Fault Tolerance Constraints. Proc. IEEE Real-Time Systems Symp., Huntsville, AL, Dec. 1988. Los Alamitos, CA: IEEE Comp. Soc. Press

NEXUS (Sect. 5.14)

419. Tripathi, A., Ghonami, A., Schmitz, T.: Object Management in the NEXUS Distributed Operating System. Proc. IEEE COMPCON Conf. Spring '87, San Francisco, CA, Feb. 1987. Los Alamitos, CA: IEEE Comp. Soc. Press, pp. 50–53

420. Tripathi, A., Ong, S.: Type Management System in the NEXUS Programming Environment. Proc. 12th IEEE Int. Computer Software and Applications Conf. COMPSAC '88, Oct. 1988. Los Alamitos, CA: IEEE Comp. Soc. Press

421. Tripathi, A.: An Overview of the NEXUS Distributed Operating System Design. IEEE Transactions on Software Engineering 15:6, 686–695 (Jun. 1989)

422. Tripathi, A.: The Object-Oriented Distributed Programming Environment of NEXUS. Dept. of Computer Science, Univ. of Minnesota, Minneapolis, MN, Technical Report TR-89-46, Jul. 1989

423. Tripathi, A., Hili, Z. Attiael: Synchronization of Nested Atomic Actions. Software – Practice and Experience 20:8, 773–798 (Aug. 1990)

PEACE   (Sect. 5.15)

424. Berg, R., Cordsen, J., Hastedt, C., Heuer, J., Nolte, J., Sander, M., Schmidt, H., Schön, F., Schröder-Preikschat, W.: Making Massively Parallel Systems Work. GMD FIRST, Berlin, Germany, Technical Report, Apr. 1990

425. Schmidt, H.: Making PEACE a Dynamic Alterable System. In: Bode, A. (ed.): Proc. 2nd Europ. Distributed Memory Computing Conf., Munich, Germany, Apr. 1991. Lecture Notes in Computer Science **487**. Berlin, Heidelberg, New York: Springer-Verlag, pp. 422–431

426. Schröder-Preikschat, W.: A Distributed Process Execution and Communication Environment for High-Performance Application Systems. In: Nehmer, J. (ed.): Proc. Int. Workshop on Experiences with Distributed Systems, Kaiserslautern, Germany, 1987. Lecture Notes in Computer Science **309**. Berlin, Heidelberg, New York: Springer-Verlag, pp. 162–188

427. Schröder, W.: PEACE: The Distributed SUPRENUM Operating System. Parallel Computing **7**, 325–333 (1988)

428. Schröder-Preikschat, W.: PEACE – A Distributed Operating System for High-Performance Multicomputer Systems. In: Schröder-Preikschat, W., Zimmer, W. (eds.): Proc. Europ. Workshop on Progress in Distributed Operating Systems and Distributed Systems Management, Berlin, Germany, Apr. 1989. Lecture Notes in Computer Science **433**. Berlin, Heidelberg, New York: Springer-Verlag, pp. 22–44

429. Schröder-Preikschat, W.: Overcoming the Startup Time Problem in Distributed Memory Architectures. Proc. 24th IEEE Hawaii Int. Conf. on System Sciences, Hawaii, HI, 1991. Los Alamitos, CA: IEEE Comp. Soc. Press

Profemo   (Sect. 5.16)

430. Kaiser, J., Nett, E., Kroeger, R.: Mutabor – A Coprocessor Supporting Object-Oriented Memory-Management and Error Recovery. Proc. 21th IEEE Hawaii Int. Conf. on System Sciences, Hawaii, HI, Jan. 1988. Los Alamitos, CA: IEEE Comp. Soc. Press

431. Mock, M.: Globales Objektmanagement in einem verteilten objektorientierten System. GMD, St. Augustin, Germany, Technical Report 128, Dec. 1987

432. Nett, E., Grosspietsch, K.E., Jungblut, A., Kaiser, J., Kroeger, R., Lux, W., Speicher, M., Winnebeck, H.W.: PROFEMO: Design and Implementation of a Fault Tolerant Distributed System Architecture. GMD, St. Augustin, Germany, Technical Report 100, Jun. 1985

433. Nett, E., Grosspietsch, K.E., Jungblut, A., Kaiser, J., Kroeger, R., Lux, W., Speicher, M., Winnebeck, H.W.: PROFEMO: Design and Implementation of a Fault Tolerant Distributed System Architecture (Formal Specification). GMD, St. Augustin, Germany, Technical Report 101, Jun. 1985

434. Nett, E., Kaiser, J., Kroeger, R.: Providing Recoverability in a Transaction Oriented Distributed Operating System. Proc 6th IEEE Int. Conf. on Distributed Computing Systems, Cambridge, MA, May 1986. Los Alamitos, CA: IEEE Comp. Soc. Press

435. Schumann, R.: Transaktions-Verwaltung in einem verteilten objektorientierten System. GMD, St. Augustin, Germany, Technical Report 134, Jan. 1988

Prospero   (Sect. 5.17)

436. Neuman, B.C.: The Virtual System Model: A Scalable Approach to Organizing Large Systems (A Thesis Proposal). Dept. of Computer Science, Univ. of Washington, Seattle, WA, Technical Report 90-05-01, May 1990

437. Neuman, B.C.: The Need for Closure in Large Distributed Systems. ACM SIGOPS Operating Systems Review **23**:4, 28–30 (Oct. 1989)

438. Neuman, B.C.: Workstations and the Virtual System Model. Proc. 2nd IEEE Workshop on Workstation Operating Systems, Pacific Grove, CA, Sep. 1989. Los Alamitos, CA: IEEE Comp. Soc. Press, pp. 91–95, also available as Univ. of Washington, Seattle, WA, Technical Report 89-10-10 and in the Newsletter of the IEEE Technical Committee on Operating Systems **3**:3, 1989

439. Neuman, B.C.: The Virtual System Model for Large Distributed Operating Systems. Dept. of Computer Science, Univ. of Washington, Seattle, WA, Technical Report 89-01-07, Apr. 1989

SOS   (Sect. 5.18)

440. Habert, S.: Gestion d'Objets et Migration dans les Systèmes Répartis. Univ. Paris VI, France, PhD thesis, Dec. 1989

441. Habert, S., Mosseri, L., Abrossimov, V.: COOL: Kernel Support for Object-Oriented Environments. Proc. Joint ECOOP/OOPSLA Conf., Ottawa, Canada, Oct. 1990

442. Makpangou, M., Shapiro, M.: The SOS Object-Oriented Communication Service. Proc. Int. Conf. on Computers and Communications, Tel Aviv, Israel, Oct. 1988

443. Makpangou, M.M.: Invocations d'objets distants dans SOS. In: Pujolle, G. (ed.): De Nouvelles Architectures pour les Communications, Paris, France, Oct. 1988. Eyrolles, pp. 195–201

444. Makpangou, M.M.: Protocoles de communication et programmation par objets : l'exemple de SOS. Univ. Paris VI, France, PhD thesis, Feb. 1989

445. Narzul, J.P. Le, Shapiro, M.: Un Service de Nommage pour un Système à Objets Répartis. Proc. AFUU Actes Convention UNIX '89, Paris, France, Mar. 1989, pp. 73–82

446. Shapiro, M.: SOS: A Distributed Object-Oriented Operating System. Proc. 2nd ACM SIGOPS Europ. Workshop on Making Distributed Systems Work, Amsterdam, the Netherlands, Sep. 1986. ACM SIGOPS Operating Systems Review **21**:1, 49–84, Jan. 1987

447. Shapiro, M.: Structure and Encapsulation in Distributed Systems: the Proxy Principle. Proc. 6th IEEE Int. Conf. on Distributed Computing Systems, Cambridge, MA, May 1986. Los Alamitos, CA: IEEE Comp. Soc. Press, pp. 198–204

448. Shapiro, M.: Structure and Encapsulation in Distributed Systems: the Proxy Principle. Proc. 6th IEEE Int. Conf. on Distributed Computing Systems, Cambridge, MA, May 1986. Los Alamitos, CA: IEEE Comp. Soc. Press, pp. 198–204

449. Shapiro, M., Abrossimov, V., Gautron, P., Habert, S., Makpangou, M.: Un Recueil de Papiers sur le Système d'Exploitation Réparti à Objets SOS. Inst. Nat. de la Recherche en Informatique et Automatique, Rocquencourt, France, Technical Report 84, May 1987

450. Shapiro, M., Gautron, P., Mosseri, L.: Persistence and Migration for C++ Objects. Proc. 3rd ECOOP '89, Nottingham, UK, Jul. 1989

451. Shapiro, M.: Prototyping a Distributed Object-Oriented OS on Unix. Proc. 1st USENIX Workshop on Experiences with Distributed and Multiprocessor Systems, Ft. Lauderdale, FL, Oct. 1989. Berkeley, CA: USENIX Association, also available as Rapport de Recherche INRIA no. 1082

452. Shapiro, M., Gourhant, Y., Habert, S., Mosseri, L., Ruffin, Mi., Valot, C.: SOS: An Object-Oriented Operating System – Assessment and Perspectives. Computing Systems 2:7 (1989)

Athena (Sect. 6.1)

453. Balkovich, E., Lerman, S., Parmelee, R.P.: Computing in Higher Education: The Athena Experience. IEEE Computer 18:11, 112–125 (Nov. 1985)

454. Balkovich, E., Lerman, S., Parmelee, R.P.: Computing in Higher Education: The Athena Experience. Communications of the ACM 28:11, 1214–1224 (Nov. 1985)

455. Champine, G.A., Geer, D.E., Ruth, W.N.: Project Athena as a Distributed Computer System. IEEE Computer 23:9, 40–51 (Sep. 1990)

456. Coppeto, T., et al.: OLC: An On-Line Consulting System for UNIX. Proc. USENIX Conf. Summer '89, Baltimore, MD, Jun. 1989. Berkeley, CA: USENIX Association, pp. 83–94

457. Gettys, J.: Project Athena. Proc. USENIX Conf. Summer '84, Salt Lake City, UT, Jun. 1984. Berkeley, CA: USENIX Association, pp. 72–77

458. Rosenstein, M., Geer, D., Levine, P.H.: The Athena Service Management System. Proc. USENIX Conf. Winter '88, Dallas, TX, Feb. 1988. Berkeley, CA: USENIX Association, pp. 203–212

459. Sonza, R.J., Miller, S.P.: UNIX and Remote Procedure Calls: A Peaceful Coexistence. Proc. 6th IEEE Int. Conf. on Distributed Computing Systems, Cambridge, MA, May 1986. Los Alamitos, CA: IEEE Comp. Soc. Press, pp. 268–277

460. Steiner, J.G., Neumann, B.C., Schiller, J.I.: Kerberos: An Authentication Service for Open Network Systems. Proc. USENIX Conf. Winter '88, Dallas, TX, Feb. 1988. Berkeley, CA: USENIX Association, pp. 191–202

461. Treese, G.W.: Berkeley UNIX on 1000 Workstations: Athena Changes to 4.3 BSD. Proc. USENIX Conf. Winter '88, Dallas, TX, Feb. 1988. Berkeley, CA: USENIX Association, pp. 175–182

Avalon (Sect. 6.2)

462. Detlefs, D.L., Herlihy, M.P., Keitzke, K.Y., Wing, J.M.: Avalon/C++: C++ Extensions for Transaction-Based Programming. Proc. USENIX Workshop on C++, Nov. 1987. Berkeley, CA: USENIX Association

463. Detlefs, D.L., Herlihy, M.P., Wing, J.M.: Inheritance of Synchronization and Recovery Properties in Avalon/C++. IEEE Computer 21:12, 57–69 (Dec. 1988)

464. Herlihy, M.P., Wing, J.M.: Avalon: Language Support for Reliable Distributed Systems. Proc. 2nd Workshop on Large-Grained Parallelism, Hidden Valley, PA, Oct. 1987, pp. 42–44, also available as Carnegie Mellon Univ., Pittsburgh, PA, Technical Report CMU-CS-86-167, July, 1987

DAPHNE (Sect. 6.3)

465. Loehr, K.-P., Mueller, J., Nentwig, L.: DAPHNE – Support for Distributed Applications Programming in Heterogeneous Computer Networks. Proc. 8th IEEE Int. Conf. on Distributed Computing Systems, San Jose, CA, Jun. 1988. Los Alamitos, CA: IEEE Comp. Soc. Press, pp. 63–71

466. Loehr, K.-P., Mueller, J., Nentwig, L.: DAPHNE – Support for Distributed Computing in Heterogeneous Environments. In: Schröder-Preikschat, W., Zimmer, W. (eds.): Proc. Europ. Workshop on Progress in Distributed Operating Systems and Distributed Systems Management, Berlin, Germany, Apr. 1989. Lecture Notes in Computer Science **433**. Berlin, Heidelberg, New York: Springer-Verlag, pp. 138–146

DASH (Sect. 6.4)

467. Anderson, D.P., Ferrari, D., Rangan, P.V., Tzou, S.-Y.: The DASH Project: Issues in the Design of a Very Large Distributed System. Computer Science Division, Univ. of California, Berkeley, CA, Technical Report UCB/CSD 87/338, 1987

468. Anderson, D.P., Ferrari, D.: The DASH Project: An Overview. Computer Science Division, Univ. of California, Berkeley, CA, Technical Report UCB/CSD 88/405, Feb. 1988

469. Anderson, D.P., Tzou, S.: The DASH Local Kernel Structure. Computer Science Division, Univ. of California, Berkeley, CA, Technical Report UCB/CSD 88/462, Nov. 1988

470. Anderson, D.P., Wahbe, R.: The DASH Network Communication Architecture. Computer Science Division, Univ. of California, Berkeley, CA, Technical Report UCB/CSD 88/462, Nov. 1988

471. Anderson, D.P.: Meta-Scheduling for Distributed Continuous Media. Computer Science Division, Univ. of California, Berkeley, CA, Technical Report UCB/CSD 90/599, Oct. 1990

472. Anderson, D.P., Herrtwich, R.G., Schaefer, C.: SRP: A Resource Reservation Protocol for Guaranteed-Performance Communication in the Internet. Int. Computer Science Inst., Berkeley, CA, Technical Report 90-006, Feb. 1990

473. Anderson, D.P., Tzou, S., Wahbe, R., Govindan, R., Andrews, M.: Support for Continuous Media in the Dash System. Proc. 10th IEEE Int. Conf. on Distributed Computing Systems, Paris, France, May 1990. Los Alamitos, CA: IEEE Comp. Soc. Press, pp. 54–61

474. Anderson, D.P., Delgrossi, L., Herrtwich, R.G.: Process Structure and Scheduling in Real-Time Protocol Implementation. In: Effelsberg, W., Meuer, H.W., Müller, G. (eds.): Proc. 7th GI/ITG Conf. – Communication in Distributed Systems, Mannheim, Germany, Feb. 1991. Informatik-Fachberichte **267**. Berlin, Heidelberg, New York: Springer-Verlag, pp. 83–95

Emerald (Sect. 6.5)

475. Black, A.P., Hutchinson, N.C., Jul, E., Levy, H.M.: Object Structure in the Emerald System. Proc. 1st ACM SIGPLAN OOPSLA '86 Conf. on Object-Oriented Programming Systems, Languages and Applications, Portland, OR, Sep. 1986. ACM SIGPLAN Notices **21**:11, pp. 78–86

476. Black, A.P., Hutchinson, N.C., Jul, E., Levy, H.M., Carter, L.: Distribution and Abstract Types in Emerald. IEEE Transactions on Software Engineering **SE–13**:1, 65–76 (Jan. 1987)

477. Hutchinson, N.C.: Emerald: A Language to Support Distributed Programming. Proc.
2nd Workshop on Large-Grained Parallelism, Hidden Valley, PA, Oct. 1987, pp. 45–47

478. Jul, E., Levy, H.M., Hutchinson, N.C., Black, A.P.: Fine-Grained Mobility in the Emerald System. ACM Transactions on Computer Systems 6:1, 109–133 (Feb. 1988)

479. Raj, R.K., Levy, H.M.: A Compositional Model for Software Reuse. Computer Journal
34:4 (Aug. 1989), also in Proc. 3rd ECOOP '89 Conf., Nottingham, UK, 1989

480. Raj, R.K., Tempero, E.D., Levy, H.M., Hutchinson, N.C., Black, A.P., Jul, E.: Emerald:
A General-Purpose Programming Language. Software – Practice and Experience 21:1,
91–118 (1991)

Enchère (Sect. 6.6)

481. Banâtre, J.P., Banâtre, M., Lapalme, G., Ployette, F.: The Design and Building of
Enchère, a Distributed Electronic Marketing System. Communications of the ACM 29:1,
19–29 (Jan. 1986)

Galaxy (Sect. 6.7)

482. Jia, X., Nakano, H., Shimizu, K., Maekawa, M.: Highly Concurrent Directory Man-
agement in the Galaxy Distributed System. Proc. 10th IEEE Int. Conf. on Distributed
Computing Systems, Paris, France, May 1990. Los Alamitos, CA: IEEE Comp. Soc.
Press, pp. 416–423

483. Shimizu, K., Maekawa, M., Hamano, J.: Hierarchical Object Groups in Distributed
Operating Systems. Proc. 8th IEEE Int. Conf. on Distributed Computing Systems, San
Jose, CA, Jun. 1988. Los Alamitos, CA: IEEE Comp. Soc. Press, pp. 18–24

484. Sinha, P.K., Ashihara, H., Shimizu, K., Maekawa, M.: Flexible User-Definable Memory
Coherence Scheme in Distributed Shared Memory of GALAXY. In: Bode, A. (ed.):
Proc. 2nd Europ. Distributed Memory Computing Conf., Munich, Germany, Apr. 1991.
Lecture Notes in Computer Science 487. Berlin, Heidelberg, New York: Springer-Verlag,
pp. 52–61

GAFFES (Sect. 6.8)

485. Gozani, S., Gray, M., Keshav, S., Madisetti, V., Munson, E., Rosenblum, M., Schoet-
tler, S., Sullivan, M., Terry, D.: GAFFES: The Design of a Globally Distributed File
System. Computer Science Division, Univ. of California, Berkeley, CA, Technical Report
UCB/CSD 87/361, Jun. 1987

486. Terry, D.B.: Distributed Name Servers: Naming and Caching in Large Distributed Computing Environments. Computer Science Division, Univ. of California, Berkeley, CA,
Technical Report UCB/CSD 85/228, Mar. 1985

Grapevine (Sect. 6.9)

487. Birrell, A.D., Levin, R., Needham, R.M., Schroeder, M.D.: Grapevine: An Exercise in
Distributed Computing. Communications of the ACM 25:4, 260–274 (Apr. 1982)

488. Schroeder, M.D., Birrell, A.D., Needham, R.M.: Experience with Grapevine: The
Growth of a Distributed System. ACM Transactions on Computer Systems 2:1, 3–23
(Feb. 1984)

HCS (Sect. 6.10)

489. Bershad, B.N., Ching, D.T, Lazowska, E.D., Sanislo, J., Schwartz, M.: A Remote Procedure Call Facility for Heterogeneous Computer Systems. IEEE Transactions on Software Engineering **SE–13**:8 (Aug. 1987)

490. Bershad, B.N., Levy, H.N.: A Remote Computation Facility for a Heterogeneous Environment. IEEE Computer **21**:5 (May 1988)

491. Black, A.P., Lazowska, E.D., Levy, H.M., Notkin, D., Sanislo, J., Zahorjan, J.: An Approach to Accommodating Heterogeneity. Dept. of Computer Science, Univ. of Washington, Seattle, WA, Technical Report 85–10–04, Oct. 1985

492. Notkin, D., Hutchinson, N., Sanislo, J., Schwartz, M.: Accomodating Heterogeneity. ACM SIGOPS Operating Systems Review **20**:2, 9–22 (Apr. 1986)

493. Notkin, D., Black, A.P., Lazowska, E.D., Levy, H.M., Sanislo, J., Zahorjan, J.: Interconnecting Heterogeneous Computer Systems. Communications of the ACM **31**:3, 258–274 (1988)

494. Pinkerton, C.B., et al.: A Heterogeneous Distributed File System. Proc. 10th IEEE Int. Conf. on Distributed Computing Systems, Paris, France, May 1990. Los Alamitos, CA: IEEE Comp. Soc. Press, pp. 424–431

495. Schwartz, M.F.: Naming in Large Heterogeneous Systems. Dept. of Computer Science, Univ. of Washington, Seattle, WA, Technical Report 87–08–01, Aug. 1987

INCAS (Sect. 6.11)

496. Nehmer, J., Beilken, C., Haban, D., Massar, R., Mattern, F., Rombach, D., Stamen, F.-J., Weitz, B., Wybranietz, D.: The Multicomputer Project INCAS – Objectives and Basic Concepts. Fachbereich Informatik, Univ. Kaiserslautern, Germany, Technical Report 11/85, Jan. 1985

497. Nehmer, J., Haban, D., Mattern, F., Wybranietz, D., Rombach, H.D.: Key Concepts of the INCAS Multicomputer Project. IEEE Transactions on Software Engineering **SE– 13**:8, 913–923 (Aug. 1987)

498. Sturm, P., Wybranietz, D., Mattern, F.: The INCAS Distributed System Project – Experience and Current Topics. Proc. Workshop Distribution and Objects, DEC Karlsruhe, Germany, 1989, pp. 97–114, also available as Technical Report 20/89, Univ. Kaiserslautern, Germany

ISIS (Sect. 6.12)

499. Birman, K.: Replication and Fault-Tolerance in the ISIS System. ACM SIGOPS Operating Systems Review **19**:5, 79–86 (Dec. 1985)

500. Birman, K., Joseph, T.: Exploiting Virtual Synchrony in Distributed Systems. Proc. 11th ACM Symp. on Operating Systems Principles, Austin, TX, Nov. 1987. ACM SIGOPS Operating Systems Review **21**:5, pp. 123–138

501. Birman, K., Joseph, T.: Reliable Communication in the Presence of Failures. ACM Transactions on Computer Systems **5**:1, 47–76 (Feb. 1987)

502. Birman, K., Joseph, T.: Exploiting replication in distributed systems. Dept. of Computer Science, Cornell Univ., Ithaca, NY, Technical Report TR 88-917, 1989

503. Birman, K., Schiper, A., Stephenson, P.: Fast Causal Multicast. Dept. of Computer Science, Cornell Univ., Ithaca, NY, Technical Report TR 90-1105, 1990

504. Joseph, T., Birman, K.: Low-Cost Management of Replicated Data in Fault-Tolerant Distributed Systems. ACM Transactions on Computer Systems 4:1, 54–70 (Feb. 1986)

505. Schmuck, F.: The Use of Efficient Broadcast Protocols in Asynchronous Distributed Systems. Dept. of Computer Science, Cornell Univ., Ithaca, NY, PhD thesis, Aug. 1988, available as Technical Report TR 88-928

Medusa (Sect. 6.13)

506. Newell, R.G.: Solid Modelling and Parametric Design in the Medusa System. Computer Graphics, pp. 223–236 (1982)

507. Newell, R.G., Parden, G.: Parametric Design in the Medusa System. Computer Applications in Production and Engineering (Apr. 1983)

508. Ousterhout, J.K., Scelza, D.A., Sindhu, P.S.: Medusa: An Experiment in Distributed Operating System Structure. Communications of the ACM 23:2, 92–105 (Feb. 1980)

509. Ousterhout, J.K.: Medusa: A Distributed Operating System. Computing Reviews 23:5 (1982)

Meglos (Sect. 6.14)

510. Ahuja, S.R.: S/NET: A High Speed Interconnect for Multiple Computers. IEEE Journal on Selected Areas in Communications SAC-1:5 (Nov. 1983)

511. Gaglianello, R.D., Katseff, H.P.: Meglos: An Operating System for a Multiprocessor Environment. Proc. 5th IEEE Int. Conf. on Distributed Computing Systems, Denver, CO, May 1985. Los Alamitos, CA: IEEE Comp. Soc. Press, pp. 35–42

512. Gaglianello, R.D., Katseff, H.P.: The Meglos User Interface. Proc. 1st ACM/IEEE AFIPS Fall Joint Computer Conf., Dallas, TX, Nov. 1986, pp. 169–177

513. Gaglianello, R.D., Katseff, H.P.: A Distributed Computing Environment for Robotics. Proc. 1986 Int. Conf. on Robotics and Automation, San Francisco, CA, Apr. 1986, pp. 1890–1896

Mirage (Sect. 6.15)

514. Fleisch, B.D.: Distributed System V in Locus: A Design and Implementation Retrospective. Proc. ACM SIGCOMM Symp. in Communications Architectures and Protocols, Stowe, VT, Aug. 1986. ACM SIGCOMM Computer Communication Review 16:3, pp. 386–396

515. Fleisch, B.D.: Distributed Shared Memory in a Loosely Coupled Distributed System. Proc. ACM SIGCOMM Workshop on Frontiers in Computer Communications Technology, Stowe, VT, Aug. 1987

516. Fleisch, B.D.: Distributed Shared Memory in a Loosely Coupled Environment. Univ. of California, Los Angeles, CA, PhD thesis, Jul. 1989

517. Fleisch, B.D., Popek, G.J.: Mirage: A Coherent Distributed Shared Memory Design. Proc. 12th ACM Symp. on Operating Systems Principles, The Wigwam Litchfield Park, AZ, Dec. 1989. ACM SIGOPS Operating Systems Review 23:5, pp. 211–223

518. Fleisch, B.D.: Reliable Distributed Shared Memory. Proc. 2nd IEEE Workshop on Experimental Distributed Systems, Huntsville, AL, Oct. 1990. Los Alamitos, CA: IEEE Comp. Soc. Press

519. Fleisch, B.D.: The Evolution of Mirage for a Workstation Environment. Tulane Univ., New Orleans, LA, Technical Report TR-90-105, Aug. 1990

520. Fleisch, B.D.: Fault Tolerant Support for Distributed Shared Memory. Proc. 4th ACM SIGOPS Europ. Workshop on Fault Tolerance Support in Distributed Systems, Bologna, Italy, Sep. 1990, also available as Tulane Univ., New Orleans, LA, Technical Report TR-90-103, June, 1990

521. Fleisch, B.D.: Replicated Data Issues for DSM. Tulane Univ., New Orleans, LA, Technical Report TR-90-104, Aug. 1990

NCA/NCS (Sect. 6.16)

522. Dineen, T.H., Leach, P.J., Mishkin, N.W., Pato, J.N., Wyant, G.L.: The Network Computing Architecture and System: An Environment for Developing Distributed Applications. Proc. USENIX Conf. Summer '87, Phoenix, AZ, Jun. 1987. Berkeley, CA: USENIX Association, pp. 385–398

523. Dineen, T.H., Leach, P.J., Mishkin, N.W., Pato, J.N., Wyant, G.L.: The Network Computing Architecture and System: An Environment for Developing Distributed Applications. Proc. IEEE COMPCON Conf. Spring '88, San Francisco, CA, Mar. 1988. Los Alamitos, CA: IEEE Comp. Soc. Press, pp. 296–299

524. Zahn, L.: Network Computing Architecture. Englewood Cliffs, NJ: Prentice-Hall, 1990

Plan9 (Sect. 6.17)

525. Presotto, D.L.: Plan 9 from Bell Labs - the Network. Proc. Europ. UNIX Systems User Group Conf. Spring '88, London, UK, Apr. 1988. Buntingford Herts, UK: EUUG, pp. 15–21

Psyche (Sect. 6.18)

526. LeBlanc, T.J., Scott, M.L., Brown, C.M.: Large-Scale Parallel Programming: Experience with the BBN Butterfly Parallel Processor. Proc. ACM SIGPLAN PPEALS – Parallel Programming: Experience with Applications, Languages and Systems, New Haven, CT, Jul. 1988. ACM SIGPLAN Notices 23:9, pp. 161–172

527. Scott, M.L., LeBlanc, T.J., Marsh, B.D.: Design Rationale for Psyche, a General-Purpose Multiprocessor Operating System. Proc. Int. Conf. on Parallel Processing, V.II – Software, Aug. 1988, pp. 255–262

528. Scott, M.L., LeBlanc, T.J., Marsh, B.D., Becker, T.G., Dubnicki, C., Markatos, E.P., Smithline, N.G.: Implementation Issues for the Psyche Multiprocessor Operating System. Computing Systems 3:1, 101–137 (Winter 1990)

529. Scott, M.L., LeBlanc, T.J., Marsh, B.D.: Multi-Model Parallel Programming in Psyche. Proc. 2nd ACM SIGPLAN PPOPP Symp. on Principles and Practice of Parallel Programming, Seattle, WA, Mar. 1990. ACM SIGPLAN Notices 25:3, pp. 70–78

SMB (Sect. 6.19)

530. Hurwicz, M.: MS-DOS 3.1 Makes it Easy to Use IBM PCs on a Network. Data Communications (Nov. 1985)

531. IBM: IBM Personal Computer Seminar Proceedings. Technical Report, May 1985

532. Stieglitz, M.: IBM Provides Industry with a Versatile Local Network Standard. Data Communications (Jun. 1985)

Symunix (Sect. 6.20)

533. Edler, J., Lipkis, J., Schonberg, E.: Process Management for Highly Parallel UNIX Systems. Proc. USENIX Workshop on UNIX and Supercomputers, Sep. 1988. Berkeley, CA: USENIX Association

534. Edler, J., Lipkis, J., Schonberg, E.: Memory Management in Symunix II: A Design for Large-Scale Shared Memory Multiprocessors. Proc. USENIX Workshop on UNIX and Supercomputers, Sep. 1988. Berkeley, CA: USENIX Association

535. Gottlieb, A., Lubachevsky, B.D., Rudolph, L.: Basic Techniques for the Efficient Coordination of Very Large Numbers of Cooperating Sequential Processors. ACM Transactions on Programming Languages and Systems 5:2, 164–189 (Apr. 1983)

536. Gottlieb, A., Grishman, R., Kruskal, C.P., McAuliffe, K.P., Rudolph, L., Snir, M.: The NYU Ultracomputer Designing a MIMD, Shared-Memory Parallel Machine. IEEE Transactions on Computers c−32:2, 175–189 (Feb. 1983)

537. Gottlieb, A.: An Overview of the NYU Ultracomputer Project. In: Dongarra, J.J. (ed.): Proc. on Experimental Parallel Computing Architectures, 1987. Amsterdam, New York: North-Holland, pp. 25–95

Synthesis (Sect. 6.21)

538. Leff, A., Pu, C., Korz, F.: Cache Performance in Server-Based and Symmetric Database Architectures. Proc. ISMM Int. Conf. on Parallel and Distributed Computing, and Systems, New York, NY, Oct. 1990

539. Massalin, H., Pu, C.: Fine-Grain Adaptive Scheduling using Feedback. Computing Systems 3:1, 139–173 (Winter 1990)

540. Massalin, H., Pu, C.: Threads and Input/Output in the Synthesis Kernel. Proc. 12th ACM Symp. on Operating Systems Principles, The Wigwam Litchfield Park, AZ, Dec. 1989. ACM SIGOPS Operating Systems Review 23:5, pp. 191–201

541. Pu, C., Massalin, H.: An Overview of the Synthesis Operating System. Dept. of Computer Science, Columbia Univ., NY, Technical Report CUCS-470-89, Oct. 1989

542. Pu, C., Massalin, H., Ioannidis, J.: The Synthesis Kernel. Computing Systems 1:1, 11–32 (1988)

543. Pu, C., Leff, A., Korz, F., Chen, S.W.: Valued Redundancy. In: Cabrera, L.-F., Pâris, J. (eds.): Proc. IEEE Workshop on Management of Replicated Data, Houston, TX, Nov. 1990. Los Alamitos, CA: IEEE Comp. Soc. Press, pp. 76–78

544. Pu, C., Chen, S.W.: Implementation of a Prototype Superdatabase. Proc. 2nd IEEE Workshop on Experimental Distributed Systems, Huntsville, AL, Oct. 1990. Los Alamitos, CA: IEEE Comp. Soc. Press

545. Pu, C., Hong, C.H., Wha, J.M.: Performance Evaluation of Global Reading of Entire Databases. Proc. Int. Symp. on Databases in Parallel and Distributed Systems, Austin, TX, Dec. 1988, pp. 167–176

546. Pu, C., Kaiser, G.E., Hutchinson, N.: Split-Transactions for Open-Ended Activities. Proc. 14th Int. Conf. on Very Large Data Bases, Los Angeles, CA, Aug. 1988. Los Altos, CA: Morgan Kaufmann Publ. Inc., pp. 27–36

547. Pu, C.: Supertransactions. Proc. 2nd Workshop on Large-Grained Parallelism, Hidden Valley, PA, Oct. 1987, pp. 72–73

x-Kernel (Sect. 6.22)

548. Hutchinson, N.C., Peterson, L.L.: The x-Kernel: An Architecture for Implementing Network Protocols. IEEE Transactions on Software Engineering SE–16 (1990)

549. Hutchinson, N.C., Mishra, S., Peterson, L.L., Thomas, V.T.: Tools for Implementing Network Protocols. Software – Practice and Experience 19:9, 895–916 (Sep. 1989)

550. Hutchinson, N.C., Peterson, L.L.: Design of the x-Kernel. Proc. 1988 Symp. on Communications Architectures and Protocols, Stanford, CA, Aug. 1988, pp. 65–75

551. Hutchinson, N.C., Peterson, L.L., Abbott, M.B., O'Malley, S.W.: RPC in the x-Kernel: Evaluating New Design Techniques. Proc. 12th ACM Symp. on Operating Systems Principles, The Wigwam Litchfield Park, AZ, Dec. 1989. ACM SIGOPS Operating Systems Review 23:5, pp. 91–101

552. Hutchinson, N.C., Peterson, L.L., O'Malley, S.W.: x-Kernel Programmer's Manual. Dept. of Computer Science, Univ. of Arizona, AZ, Technical Report 89-29, Dec. 1989

553. O'Malley, S.W., Abbott, M., Hutchinson, N.C., Peterson, L.L.: A Transparent Blast Facility. Journal of Internetworking 1:1 (Sep. 1990)

554. O'Malley, S.W., Peterson, L.L.: A Highly Layered Architecture for High-Speed Networks. Dept. of Computer Science, Univ. of Arizona, AZ, Technical Report 90-22, Jun. 1990

555. Peterson, L.L., Hutchinson, N.C., O'Malley, S.W., Rao, H.: The x-Kernel: A Platform for Accessing Internet Resources. IEEE Computer 23:5, 23–33 (May 1990)

Acorn (Sect. 8.1)

556. Dellar, C.: A File Server for a Network of Low Cost Personal Microcomputers. Software – Practice and Experience 12:12, 1051–1068 (Dec. 1982)

Agora (Sect. 8.2)

557. Bisiani, R., Forim, A.: Multilanguage Parallel Programming. Proc. Int. Conf. on Parallel Processing, Aug. 1987, pp. 381–384

558. Bisiani, R., Forin, A.: Agora: Heterogeneous and Multilanguage Parallel Programming. Proc. 2nd Workshop on Large-Grained Parallelism, Hidden Valley, PA, Oct. 1987, pp. 8–10

Amber (Sect. 8.3)

559. Chase, J.S., Amador, F.G., Lazowska, E.D., Levy, H.M., Littlefield, R.J.: The Amber System: Parallel Programming on a Network of Multiprocessors. Proc. 12th ACM Symp. on Operating Systems Principles, The Wigwam Litchfield Park, AZ, Dec. 1989. ACM SIGOPS Operating Systems Review 23:5, pp. 147–158

Arachne (Sect. 8.4)

560. Finkel, R.A., Solomon, M.H.: The Arachne Kernel, Version 1.2. Dept. of Computer Science, Univ. of Wisconsin-Madison, Madison, WI, Technical Report 380, Apr. 1980

561. Finkel, R.A., Solomon, M.H.: The Arachne Distributed Operating System. Dept. of Computer Science, Univ. of Wisconsin-Madison, Madison, WI, Technical Report 439, Jul. 1981

Arca (Sect. 8.5)

562. Muir, S., Hutchison, D., Shepherd, D.: Arca: A Local Network File Server. Computer Journal **28**:3, 243–249 (Mar. 1985)

Arcade (Sect. 8.6)

563. Cohn, D.L., Delaney, W.P., Tracey, K.M.: Arcade: An Architecture for a Distributed Environment. Dept. of Electrical and Computer Engineering, Univ. of Notre Dame, Notre Dame, IN, Technical Report 889, Nov. 1988

564. Cohn, D.L., Delaney, W.P., Tracey, K.M.: Arcade: A Platform for Heterogeneous Distributed Operating Systems. Proc. 1st USENIX Workshop on Experiences with Distributed and Multiprocessor Systems, Ft. Lauderdale, FL, Oct. 1989. Berkeley, CA: USENIX Association, pp. 373–390

565. Cohn, D.L., Delaney, W.P., Tracey, K.M.: Structured Shared Memory Among Heterogeneous Machines in Arcade. Proc. 1st IEEE Symp. on Parallel and Distributed Processing, May 1989. Los Alamitos, CA: IEEE Comp. Soc. Press, pp. 378–379

566. Delaney, W.P.: The Arcade Distributed Environment: Design, Implementation and Analysis. Dept. of Electrical and Computer Engineering, Univ. of Notre Dame, Notre Dame, IN, PhD thesis, Apr. 1989

Archons (Sect. 8.7)

567. Colwell, R.P., Hitchcock, C.Y., Jensen, E.D., Kollar, C., Sprunt, B.: Computers, Complexity, and Controversy. IEEE Computer **18**:9 (Sep. 1985)

568. Jensen, E.D., Pleszkoch, N.L.: ArchOS–A Physically Dispersed Operating System: An Overview of its Objectives and Approach. IEEE Distributed Processing Technical Committee Newsletter – Special Issue on Distributed Operating systems (Jun. 1984)

569. Jensen, E.D.: ArchOS. Proc. ACM SIGOPS Workshop on Operating Systems in Computer Networks, Rüschlikon, Switzerland, Jan. 1985. ACM SIGOPS Operating Systems Review **19**:2, 6–40, Apr. 1985

570. Sha, L., Jensen, E.D., Rashid, R., Northcutt, J.D.: Distributed Cooperating Processes and Transactions. Proc. ACM Conf. on Data Communication Protocols and Architectures, Mar. 1983

571. Sha, L., Lehoczky, J.P., Jensen, E.D.: Modular Concurrency Control and Failure Recovery. IEEE Transactions on Computers **c–37**:2 (Feb. 1988)

Argos (Sect. 8.8)

572. Johnson, E.E.: Gmmp Multiprocessor Architectures. Proc. Int. Conf. on Computing and Information **1**, 1989. Amsterdam, New York: North-Holland

573. Johnson, E.E.: Argos – A Research Gmmp Operating System. Dept. of Electrical and Computer Engineering, New Mexico State Univ., Las Cruces, NM, Technical Report, Aug. 1989

574. Johnson, E.E.: A Multiprocessor Architecture for High-Rate Communication Processing. Proc. MILCOM '91, 1991

Arjuna   (Sect. 8.9)

575. Dixon, G.N., Shrivastava, S.K., Parrington, G.D.: Managing Persistent Objects in Arjuna: A System for Reliable Distributed Computing. Proc. Workshop on Persistent Object System: Their Design, Implementation and Use, Appin, UK, Aug. 1987

576. Dixon, G.N., Parrington, G.D., Shrivastava, S.K., Wheater, S.M.: The Treatment of Persistent Objects in Arjuna. Computer Journal **32**:4, 323–332 (Aug. 1989)

577. Little, M.C., Shrivastava, S.K.: Replicated K-Resilient Objects in Arjuna. In: Cabrera, L.-F., Pâris, J. (eds.): Proc. IEEE Workshop on Management of Replicated Data, Houston, TX, Nov. 1990. Los Alamitos, CA: IEEE Comp. Soc. Press, pp. 53–58

BCIS   (Sect. 8.10)

578. Gifford, D.K.: Polychannel Systems for Mass Digital Communication. Communications of the ACM **33**:2, 141–151 (Feb. 1990)

Camelot   (Sect. 8.11)

579. Spector, A.Z., Bloch, J.J., Daniels, D.S., Draves, R.P., Duchamp, D., Eppinger, J.L., Menees, S.G., Thompson, D.S.: The Camelot Project. Dept. of Computer Science, Carnegie Mellon Univ., Pittsburgh, PA, Technical Report CMU–CS–86–166, 1986

580. Spector, A.Z.: Distributed Transaction Processing and the Camelot System. Dept. of Computer Science, Carnegie Mellon Univ., Pittsburgh, PA, Technical Report CMU–CS–87–100, 1987

581. Spector, A.Z., Thompson, D.S., Pausch, R.F., Eppinger, J.L., Duchamp, D., Draves, R.P., Daniels, D.S., Bloch, J.J.: Camelot: A Distributed Transaction Facility for Mach and the Internet. Dept. of Computer Science, Carnegie Mellon Univ., Pittsburgh, PA, Technical Report CMU–CS–87–129, 1987

582. Spector, A.Z.: Camelot: A Flexible, Distributed Transaction Processing System. Proc. IEEE COMPCON Conf. Spring '88, San Francisco, CA, Mar. 1988. Los Alamitos, CA: IEEE Comp. Soc. Press, pp. 432–436

CMCFS   (Sect. 8.12)

583. Ball, E.J., Barbacci, M.R., Fahlman, S.E., Harbison, S.P., Hibbard, P.G., Rashid, R.F., Robertson, G.G., Jr., G.L. Steele: The Spice Project. Computer Science Research Review 1980–1981, pp. 5–36 (1982), Dept. of Computer Science, Carnegie Mellon Univ., Pittsburgh, PA

Choices   (Sect. 8.13)

584. Campbell, R.H., Johnston, G.M., Madany, P.W., Russo, V.F.: Principles of Object-Oriented Operating System Design. Dept. of Computer Science, Univ. of Illinois, Urbana, IL, Technical Report R–89–1510, Apr. 1989

Circus   (Sect. 8.14)

585. Cooper, E.C.: Replicated Distributed Programs. Computer Science Division, Univ. of California, Berkeley, CA, Technical Report UCB/CSD 85/231, May 1985

586. Cooper, E.C.: Replicated Distributed Programs. ACM SIGOPS Operating Systems Review **19**:5, 63–78 (Dec. 1985)

Clearinghouse (Sect. 8.15)

587. Oppen, D.C., Dalal, Y.K.: The Clearinghouse: A Decentralized Agent for Locating Named Objects in a Distributed Environment. ACM Transactions on Office Information Systems 1:3, 230–253 (Jul. 1983)

Cocanet (Sect. 8.16)

588. Rowe, L.A., Birman, K.P.: A Local Network Based on the UNIX Operating System. IEEE Transactions on Software Engineering **SE–8**:2, 137–146 (Mar. 1982)

CSSA (Sect. 8.17)

589. Mattern, F., Beilken, C.: The Distributed Programming Language CSSA – A very short Introduction. Fachbereich Informatik, Univ. Kaiserslautern, Germany, Technical Report 123/85, Jan. 1985

590. Mattern, F.: CSSA – Sprache und Systembenutzung. Fachbereich Informatik, Univ. Kaiserslautern, Germany, Technical Report 24/88, Mar. 1988

CONIC (Sect. 8.18)

591. Magee, J., Kramer, J., Sloman, M.: The CONIC Support Environment for Distributed Systems. In: Paker, Y., Banatre, J.-P., Bozyigit, M. (eds.): Distributed Operating Systems: Theory and Practice, NATO ASI series Vol. F28, 1987. Berlin, Heidelberg, New York: Springer-Verlag, pp. 289–311

CICS (Sect. 8.19)

592. Yelavich, M.B.: Customer Information Control System – An Evolving System Facility. IBM Systems Journal **24**:4, 263–278 (1985)

Datacomputer (Sect. 8.20)

593. Marill, T., Stern, D.: The Datacomputer – A Network Data Utility. Proc. AFIPS NCC **44**, 1975, pp. 389–395

Deceit (Sect. 8.21)

594. Siegel, A., Birman, K., Marzullo, K.: Deceit: A Flexible Distributed File System. Proc. USENIX Conf. Summer '90, Anaheim, CA, Jun. 1990. Berkeley, CA: USENIX Association, also in: L.-F. Cabrera and J. Pâris (eds.): Proc. IEEE Workshop on Management of Replicated Data, Houston, TX, pp. 15–17, Nov. 1990

DEMOS (Sect. 8.22)

595. Baskett, F., Howard, J.H., Montague, J.T.: Task Communications in DEMOS. Proc. 6th ACM Symp. on Operating Systems Principles, Purdue Univ., West Lafayette, IN, Nov. 1977. ACM SIGOPS Operating Systems Review **11**:5, pp. 23–32

DFS925 (Sect. 8.23)

596. Arden, M.J.: DFS925: A Distributed File System in a Workstation/LAN Environment. Computer Science Division, Univ. of California, Berkeley, CA, Technical Report UCB/CSD 85/236, May 1985

DISTRIX (Sect. 8.24)

597. Christie, D.: DISTRIX. Proc. ACM SIGOPS Workshop on Operating Systems in Computer Networks, Rüschlikon, Switzerland, Jan. 1985. ACM SIGOPS Operating Systems Review 19:2, 6–40, Apr. 1985

Dragon-Slayer (Sect. 8.25)

598. Daniels, D.C.: Dragon Slayer: A Blueprint for the Design of a Completely Distributed Operating System. Wayne State Univ., Detroit, MI, Master's thesis, 1986

Echo (Sect. 8.26)

599. Hisgen, A., Birrel, A., Mann, T., Schroeder, M., Swart, G.: Availability and Consistency Trade-Offs in the Echo Distributed File System. Proc. 2nd IEEE Workshop on Workstation Operating Systems, Pacific Grove, CA, Sep. 1989. Los Alamitos, CA: IEEE Comp. Soc. Press

600. Hisgen, A., Birrel, A., Jerian, C., Mann, T., Schroeder, M., Swart, G.: Granularity and Semantic Level of Replication in the Echo Distributed File System. In: Cabrera, L.-F., Pâris, J. (eds.): Proc. IEEE Workshop on Management of Replicated Data, Houston, TX, Nov. 1990. Los Alamitos, CA: IEEE Comp. Soc. Press, pp. 2–4

601. Mann, T., Hisgen, A., Swart, G.: An Algorithm for Data Replication. DEC Systems Research Center, Palo Alto, CA, Technical Report 46, Jun. 1989

Encompass (Sect. 8.27)

602. Borr, A.: Transaction Monitoring in ENCOMPASS: Reliable Distributed Transaction Processing. Proc. 7th Int. Conf. on Very Large Data Bases, Cannes, France, Sep. 1981. Los Altos, CA: Morgan Kaufmann Publ. Inc., also TANDEM TR 81.2, Tandem Computers Inc., Cupertino, CA, June 1981

Felix (Sect. 8.28)

603. Fridrich, M., Older, W.: The Felix File Server. Proc. 8th ACM Symp. on Operating Systems Principles, Asilomar, CA, Dec. 1981. ACM SIGOPS Operating Systems Review 15:5, pp. 37–44

Ficus (Sect. 8.29)

604. Guy, R.G., Heidemann, J.S., Mak, W., Page, T.W., Popek, G.J., Rothmeier, D.: Implementation of the Ficus Replicated File System. Proc. USENIX Conf. Summer '90, Anaheim, CA, Jun. 1990. Berkeley, CA: USENIX Association, pp. 63–71

605. Guy, R.G., Page, T.W., Heidemann, J.S., Popek, G.J.: Name Transparency in Very Large Scale Distributed File Systems. Proc. 2nd IEEE Workshop on Experimental Distributed Systems, Huntsville, AL, Oct. 1990. Los Alamitos, CA: IEEE Comp. Soc. Press

606. Guy, R.G.: Ficus: A Very Large Scale Reliable Distributed File System. Dept. of Computer Science, Univ. of California, Los Angeles, CA, PhD thesis, Jun. 1991, also available as Univ. of California, Los Angeles, CA, Technical Report CSD-910018

607. Heidemann, J.S., Popek, G.J.: An Extensible, Stackable Method of File System Development. Univ. of California, Los Angeles, CA, Technical Report CSD-900044, Dec. 1990.

608. Page, T.W., Popek, G.J., Guy, R.G., Heidemann, J.S.: The Ficus Distributed File System: Replication via Stackable Layers. Univ. of California, Los Angeles, CA, Technical
Report CSD-900009, Apr. 1990

609. Page, T.W., Guy, R.G., Popek, G.J., Heidemann, J.S.: Architecture of the Ficus Scalable
Replicated File System. Univ. of California, Los Angeles, CA, Technical Report CSD-
910005, Mar. 1991

610. Popek, G.J., Guy, R.G., Page, T.W., Heidemann, J.S.: Replication in Ficus Distributed
File System. In: Cabrera, L.-F., Pâris, J. (eds.): Proc. IEEE Workshop on Management
of Replicated Data, Houston, TX, Nov. 1990. Los Alamitos, CA: IEEE Comp. Soc.
Press, pp. 5–10

FileNet   (Sect. 8.30)

611. Edwards, D.A., McKendry, M.S.: Exploiting Read-Mostly Workloads in the FileNet System. Proc. 12th ACM Symp. on Operating Systems Principles, The Wigwam Litchfield
Park, AZ, Dec. 1989. ACM SIGOPS Operating Systems Review **23**:5, pp. 58–70

612. McKendry, M.S.: The FileNet System. Proc. 2nd Workshop on Large-Grained Parallelism, Hidden Valley, PA, Oct. 1987, pp. 60–62

Firefly   (Sect. 8.31)

613. Thacker, C.P., Stewart, L.C.: Firefly: A Multiprocessor Workstation. Proc. Int. Conf. on
Architectural Support for Programming Languages and Operating Systems, Palo Alto,
CA, Feb. 1987, pp. 164–172

614. Schroeder, M.D., Burrows, M.: Performance of Firefly RPC. Proc. 12th ACM Symp.
on Operating Systems Principles, The Wigwam Litchfield Park, AZ, Dec. 1989. ACM
SIGOPS Operating Systems Review **23**:5, pp. 83–90

GFS   (Sect. 8.32)

615. Rodriguez, R., Koehler, M., Hyde, R.: The Generic File System. Proc. Europ. UNIX
Systems User Group Conf. Autumn '86, Manchester, UK, Sep. 1986. Buntingford Herts,
UK: EUUG, pp. 83–92

Helios   (Sect. 8.33)

616. Schabernack, J., Schütte, A.: Helios, a Distributed Operating System for Transputer
Systems. Informationstechnik **32**:2, 107–114 (1990), (in German)

617. Joosen, W., Sneyers, M., Berbers, Y., Verbaeten, P.: Evaluating Communication Overhead in Helios. Proc. 4th Conf. of the North American Transputer Users Group, Ithaca,
NY, 1990

618. Perihelion Software, ltd.: The Helios Operating System. Englewood Cliffs, NJ: Prentice-
Hall, 1989

HERON   (Sect. 8.34)

619. private communication

IDRPS   (Sect. 8.35)

620. Carson, J.H.: A Distributed Operating System for a Workstation Environment. Proc.
7th Phoenix Conf. on Computers and Communication, Scottsdale, AZ, Mar. 1988. Los
Alamitos, CA: IEEE Comp. Soc. Press, pp. 213–217

IFS   (Sect. 8.36)

621. Thacker, C.P., McCreight, E.M., Lampson, B.W., Sprol, R.F., Boogs, D.R.: Alto: A Personal Computer. In: Siewiorek, D.P., Bell, C.G., Newell, A. (eds.): Computer Structures:
Principles and Examples, 1981. McGraw-Hill, New York

LADY   (Sect. 8.37)

622. Haban, D.: Description of the System Implementation Language LADY. Fachbereich
Informatik, Univ. Kaiserslautern, Germany, Technical Report 29/85, Oct. 1985

623. Wybranietz, D., Massar, R.: An Overview of LADY – A Language for the Implementation of Distributed Operating Systems. Fachbereich Informatik, Univ. Kaiserslautern,
Germany, Technical Report 12/85, Jan. 1985

624. Wybranietz, D., Haban, D., Buhler, P.: Some Extensions of the LADY language. Fachbereich Informatik, Univ. Kaiserslautern, Germany, Technical Report 28/86, Oct. 1986

625. Wybranietz, D., Buhler, P.: The LADY Programming Environment for Distributed Operating Systems. Future Generation Computer Systems Journal **6**:3, 209–223 (1990)

Lynx   (Sect. 8.38)

626. Scott, M.L.: Language Support for Loosely-Coupled Distributed Programs. IEEE Transactions on Software Engineering **SE–13**:1, 88–103 (Jan. 1987)

627. Scott, M.L., Cox, A.L.: An Empirical Study of Message-Passing Overhead. Proc. 7th
IEEE Int. Conf. on Distributed Computing Systems, Berlin, Germany, Sep. 1987. Los
Alamitos, CA: IEEE Comp. Soc. Press, pp. 536–543

MANDIS   (Sect. 8.39)

628. Holden, D., Langsford, A.: MANDIS: Management of Distributed Systems. In: Schröder-Preikschat, W., Zimmer, W. (eds.): Proc. Europ. Workshop on Progress in Distributed
Operating Systems and Distributed Systems Management, Berlin, Germany, Apr. 1989.
Lecture Notes in Computer Science **433**. Berlin, Heidelberg, New York: Springer-Verlag,
pp. 162–173

629. Langsford, A.: MANDIS: An Experiment in Distributed Processing. In: Speth, R. (ed.):
Proc. EUTECO '88 – Research into Networks and Distributed Applications, Vienna,
Austria, Apr. 1988. Amsterdam, New York: North-Holland, pp. 787–794

Melampus   (Sect. 8.40)

630. Cabrera, L.-F., Haas, L., Richardson, J., Schwarz, P., Stamos, J.: The Melampus Project:
Toward an Omniscient Computing System. IBM Almaden Research Center, San Jose,
CA, Technical Report RJ7515, Jun. 1990

631. Cabrera, L.-F., Schwarz, P.: Operating System Support for an Omniscient Information
System. Proc. OOPSLA/ECOOP '90 Workshop on Object Orientation in Operating
Systems, Ottawa, Canada, Oct. 1990

632. Richardson, J., Schwarz, P.: Aspects: Extending Objects to Support Multiple, Independent Roles. IBM Almaden Research Center, San Jose, CA, Technical Report RJ7657, Aug. 1990

Meta (Sect. 8.41)

633. Marzullo, K., Wood, M., Cooper, R., Birman, K.: Tools for Distributed Application Management. Dept. of Computer Science, Cornell Univ., Ithaca, NY, Technical Report TR 90-1136, 1990

MICROS (Sect. 8.42)

634. Wittie, L.D., Tilborg, A.M.van: MICROS, a Distributed Operating System for MICRONET, a Reconfigurable Network Computer. IEEE Transactions on Computers c–29:12, 1133–1144 (Dec. 1980)

MODOS (Sect. 8.43)

635. Autenrieth, K., Dappa, H., Grevel, M., Kubalski, W., Bartsch, T.: Konzeption für ein verteiltes echtzeitfähiges Betriebssystem für Kommunikationssysteme. Telenorma BOSCH Telecom, Germany, Technical Report, 1989

636. Autenrieth, K., Dappa, H., Grevel, M., Kubalski, W., Bartsch, T.: Konzepte verteilter Betriebssysteme. Heidelberg: Hüthig Verlag, 1990

Munin (Sect. 8.44)

637. Bennett, J.K., Carter, J.B., Zwaenepoel, W.: Munin: Distributed Shared Memory Based on Type-Specific Memory Coherence. Proc. 2nd ACM SIGPLAN PPOPP Symp. on Principles and Practice of Parallel Programming, Seattle, WA, Mar. 1990. ACM SIGPLAN Notices 25:3, pp. 168–176

638. Bennett, J.K., Carter, J.B., Zwaenepoel, W.: Adaptive Software Cache Management for Distributed Shared Memory Architecture. Proc. 7th Int. Symp. on Computer Architecture, May 1990, pp. 125–134, also available as Rice Univ., Technical Report RICE-COMP TR90-109

NEST (Sect. 8.45)

639. Agrawal, R., Ezzat, A.K.: Processor Sharing in NEST: A Network of Computer Workstations. Proc. 1st Int. Conf. on Computer Workstations, San Jose, CA, Nov. 1985, pp. 198–208

640. Agrawal, R.: Using a Network of Computer Workstations as a Loosely-Coupled Multiprocessor. Proc. 2nd Workshop on Large-Grained Parallelism, Hidden Valley, PA, Oct. 1987, p. 1

641. Agrawal, R., Ezzat, A.K.: Location Independent Remote Execution in NEST. IEEE Transactions on Software Engineering SE–13:8, 905–912 (Aug. 1987)

642. Ezzat, A.K.: Load Balancing in NEST: A Network of Computer Workstations. Proc. 1st ACM/IEEE AFIPS Fall Joint Computer Conf., Dallas, TX, Nov. 1986, pp. 1138–1148

NRDP (Sect. 8.46)

643. Schwartz, M.F., Tsirigotis, P.G.: Experience with a Semantically Cognizant Internet White Pages Directory Tool. Journal of Internetworking Research and Experience, pp. 23–50 (1988)

644. Schwartz, M.F.: The Networked Resource Discovery Project. Proc. IFIP XI World Congress, San Francisco, CA, 1989, pp. 827–832

645. Schwartz, M.F.: A Scalable, Non-Hierarchical Resource Discovery Mechanism Based on Probabilistic Protocols. Dept. of Computer Science, Univ. of Colorado, Boulder, CO, Technical Report CU–CS–474–90, 1990

646. Schwartz, M.F., Wood, D.C.M.: A Measurement Study of Organizational Properties in the Global Electronic Mail Community. Dept. of Computer Science, Univ. of Colorado, Boulder, CO, Technical Report CU–CS–482–90, 1990

647. Schwartz, M.F., Hardy, D.R., Heinzman, W.K., Hirschowitz, G.: Supporting Resource Discovery Among Public Internet Archives Using a Spectrum of Information Quality. Proc. 11th IEEE Int. Conf. on Distributed Computing Systems, Arlington, TX, May 1991. Los Alamitos, CA: IEEE Comp. Soc. Press, pp. 82–89

648. Schwartz, M.F., Tsirigotis, P.G.: Techniques for Supporting Wide Area Distributed Applications. Dept. of Computer Science, Univ. of Colorado, Boulder, CO, Technical Report CU–CS–519–91, 1991

649. Schwartz, M.F., Goldstein, D.H., Neves, R.K., Wood, D.C.M.: An Architecture for Discovering and Visualizing Characteristics of Large Internets. Dept. of Computer Science, Univ. of Colorado, Boulder, CO, Technical Report CU–CS–520–91, 1991

650. Schwartz, M.F.: The Great Disconnection. Dept. of Computer Science, Univ. of Colorado, Boulder, CO, Technical Report CU–CS–521–91, 1991

NonStop (Sect. 8.47)

651. Bartlett, J.F.: A NonStop Kernel. Proc. 8th ACM Symp. on Operating Systems Principles, Asilomar, CA, Dec. 1981. ACM SIGOPS Operating Systems Review 15:5, pp. 22–29

Onyx (Sect. 8.48)

652. Tripathi, A., Johnson, S., Clark, R.: Onyx: An Object-Oriented Distributed Programming Language for NEXUS. Dept. of Computer Science, Univ. of Minnesota, Minneapolis, MN, Technical Report TR-89-68, Oct. 1989

PHARROS (Sect. 8.49)

653. Zandt, J.van: The PHARROS project. Proc. 2nd Workshop on Large-Grained Parallelism, Hidden Valley, PA, Oct. 1987, p. 86

Presto (Sect. 8.50)

654. Bershad, B.N., Lazowska, E.D., Levy, H.M.: PRESTO: A System for Object-Oriented Parallel Programming. Software – Practice and Experience 18:8, 713–732 (Aug. 1988)

R⋆   (Sect. 8.51)

655. Lindsay, B.G., Haas, L.M., Mohan, C., Wilms, P.F., Yost, R.A.: Computation and Communication in R⋆: A Distributed Database Manager. ACM Transactions on Computer Systems **2**:1, 24–38 (Feb. 1984)

Rapport   (Sect. 8.52)

656. Ahuja, S.R., Ensor, J.R., Horn, D.N.: Parallelism in the Rapport Multimedia Conferencing System. Proc. 2nd Workshop on Large-Grained Parallelism, Hidden Valley, PA, Oct. 1987, pp. 2–3

657. Ahuja, S.R., Ensor, J.R., Horn, D.N.: The Rapport Multimedia Conferencing System. Proc. Conf. on Office Information Systems, Palo Alto, CA, Mar. 1988, pp. 1–8

RNFS   (Sect. 8.53)

658. Marzullo, K., Schmuck, F.: Supplying high Availability with a Standard Network File System. Proc. 8th IEEE Int. Conf. on Distributed Computing Systems, San Jose, CA, Jun. 1988. Los Alamitos, CA: IEEE Comp. Soc. Press, pp. 447–453

RIG   (Sect. 8.54)

659. Ball, J.E., Feldman, J.R., Low, J.R., Rashid, R.F., Rovner, P.D.: RIG, Rochester's Intelligent Gateway: System Overview. IEEE Transactions on Software Engineering **SE–2**:4, 321–328 (Dec. 1976)

ROE   (Sect. 8.55)

660. Ellis, C.S., Floyd, R.A.: The ROE file system. Proc. 3rd IEEE Symp. on Reliability in Distributed Software and Database Systems, Clearwater Beach, FL, Oct. 1983. Los Alamitos, CA: IEEE Comp. Soc. Press, pp. 17–19

661. Floyd, R.A.: Transparency in Distributed File Systems. Dept. of Computer Science, Univ. of Rochester, NY, Technical Report 272, 1989

Roscoe   (Sect. 8.56)

662. Finkel, R.A., Solomon, M.H.: The Roscoe Kernel. Dept. of Computer Science, Univ. of Wisconsin-Madison, Madison, WI, Technical Report 337, Oct. 1978

663. Solomon, M.H., Finkel, R.A.: Roscoe: A Multi-Microcomputer Operating System. Dept. of Computer Science, Univ. of Wisconsin-Madison, Madison, WI, Technical Report 321, Apr. 1978

664. Tischler, R., Solomon, M.H., Finkel, R.A.: Roscoe Users Guide. Dept. of Computer Science, Univ. of Wisconsin-Madison, Madison, WI, Technical Report 336, Oct. 1978

665. Tischler, R., Finkel, R.A., Solomon, M.H.: Roscoe Utility Processes. Dept. of Computer Science, Univ. of Wisconsin-Madison, Madison, WI, Technical Report 338, Oct. 1978

RSS   (Sect. 8.57)

666. Gray, J.N., McJones, P., Blasgen, M.W., Lorie, R.A., Price, T.G., Putzulu, G.F., Traiger, I.L.: The Recovery Manager of the System R Database Manager. ACM Computing Surveys **13**:2, 223–242 (Jun. 1981)

RTPCDS   (Sect. 8.58)

667. Sauer, C.H., Johnson, D.W., Loucks, L.K., Shaheen-Gouda, A.A., Smith, T.A.: RT PC Distributed Services Overview. ACM SIGOPS Operating Systems Review **21**:3, 18–29 (Jul. 1987)

Linda   (Sect. 8.59)

668. Carriero, N., Gelernter, D.: The S/Net's Linda Kernel. ACM Transactions on Computer Systems **4**:2, 110–129 (May 1986)

669. Gelernter, D.: Generative Communication in Linda. ACM Transactions on Programming Languages and Systems **7**:1, 80–112 (Jan. 1985)

670. Gelernter, D., Carriero, N., Chandran, S., Chang, S.: Parallel Programming in Linda. Proc. Int. Conf. on Parallel Processing, Aug. 1985, pp. 255–263

Sesame   (Sect. 8.60)

671. Thompson, M.R., Sansom, R.D., Jones, M.B., Rashid, R.F.: Sesame: The Spice File System. Dept. of Computer Science, Carnegie Mellon Univ., Pittsburgh, PA, Technical Report CMU–CS–85–172, Dec. 1985

StarOS   (Sect. 8.61)

672. Levy, H.M.: The StarOS System. Capability-based Computer Systems, 1984. Bedford, MA: Digital Press, pp. 127–137

STORK   (Sect. 8.62)

673. Pâris, J.-F., Tichy, W.F.: STORK: An Experimental Migrating File System for Computer Networks. Proc. IEEE INFOCOM, San Diego, CA, Apr. 1983. Los Alamitos, CA: IEEE Comp. Soc. Press, pp. 168–175

Thoth   (Sect. 8.63)

674. Cheriton, D.R., Malcolm, M.A., Melen, L.S., Sager, G.R.: Thoth, a Portable Real-Time Operating System. Communications of the ACM **22**:2, 105–115 (Feb. 1979)

675. Cheriton, D.R.: The Thoth System: Multi-Process Structuring and Portability. Amsterdam, New York: North-Holland, 1982

TimixV2   (Sect. 8.64)

676. Abdallah, H.B.: Design and Implementation of User Datagram/Internet Protocol over TimixV2. Dept. of Computer and Information Science, Univ. of Pennsylvania, Philadelphia, PA, Master's thesis, May 1991

677. King, R.B.: Design, Implementation, and Evaluation of a Distributed Real-Time Kernel for Distributed Robotics. Dept. of Computer and Information Science, Univ. of Pennsylvania, Philadelphia, PA, Technical Report MS–CIS–90–40/GRASP LAB 220, Jul. 1990

678. Lee, I., King, R.B.: Timix: A Distributed Real-Time Kernel for Multi-Sensor Robots. Proc. IEEE Int. Conf. on Robotics and Automation, Apr. 1988. IEEE Council on Robotics and Automation, Los Alamitos, CA: IEEE Comp. Soc. Press, pp. 1587–1589

679. Lee, I., King, R.B., Paul, R.P.: A Predictable Real-Time Kernel for Distributed Multi-Sensor Systems. IEEE Computer **22**:6, 78–83 (Jun. 1988)

Topaz (Sect. 8.65)

680. McJones, P., Hisgen, A.: The Topaz System: Distributed Multiprocessor Personal Computing. Proc. 1st IEEE Workshop on Workstation Operating Systems, Cambridge, MA, Nov. 1987. Los Alamitos, CA: IEEE Comp. Soc. Press

TILDE (Sect. 8.66)

681. Comer, D.: Transparent Integrated Local and Distributed Environment (TILDE) Project Overview. Dept. of Computer Science, Purdue Univ., West Lafayette, IN, Technical Report CSD-TR-466, 1984

TRFS (Sect. 8.67)

682. Hughes, R.P.: The Transparent Remote File System. Proc. USENIX Conf. Summer '86, Atlanta, GA, Jun. 1986. Berkeley, CA: USENIX Association, pp. 306–317

Trollius (Sect. 8.68)

683. Burns, G.D.: Trollius: Early American Transputer Software. Ohio State Univ., Columbus, OH, (1988)

UPPER (Sect. 8.69)

684. Giloi, W.G., Behr, P.M.: Obtaining a Secure, Fault-Tolerant Distributed System with Maximal Performance. Proc. IFIP Workshop on Hardware Supported Implementation of Concurrent Languages in Distributed Systems, Bristol, UK, 1984, pp. 101–114

WFS (Sect. 8.70)

685. Swinehart, D.C., McDaniel, G., Boggs, D.: WFS: A Simple Shared File System for a Distributed Environment. Proc. 7th ACM Symp. on Operating Systems Principles, Pacific Grove, CA, Dec. 1979. ACM SIGOPS Operating Systems Review 13:5, pp. 9–17

Xcode (Sect. 8.71)

686. Tisato, F., Zicari, R.: The Xcode Machine. Proc. 3rd Symp. on Microcomputer and Microprocessor Applications, Budapest, Hungary, Oct. 1983. ACM SIGSMALL Newsletter 10:1, 1984

Z-Ring (Sect. 8.72)

687. Bux, W., Closs, F., Janson, P., Kuemmerle, K., Mueller, H.R., Rothhauser, E.H.: A Local-Area Communication Network Based on a Reliable Token-Ring System. In: Ravasio, P.C., Hopkins, G., Naffah, N. (eds.): Proc. Int. Symp. on Local Computer Networks, Florence, Italy, Apr. 1982. Amsterdam, New York: North-Holland, pp. 69–82